The Role of Organic Matter in Structuring Microbial Communities

The mission of the Awwa Research Foundation (AwwaRF) is to advance the science of water to improve the quality of life. Funded primarily through annual subscription payments from over 1,000 utilities, consulting firms, and manufacturers in North America and abroad, AwwaRF sponsors research on all aspects of drinking water, including supply and resources, treatment, monitoring and analysis, distribution, management, and health effects.

From its headquarters in Denver, Colorado, the AwwaRF staff directs and supports the efforts of over 800 volunteers, who are the heart of the research program. These volunteers, serving on various boards and committees, use their expertise to select and monitor research studies to benefit the entire drinking water community.

Research findings are disseminated through a number of technology transfer activities, including research reports, conferences, videotape summaries, and periodicals.

The Role of Organic Matter in Structuring Microbial Communities

Prepared by:
Louis A. Kaplan
Stroud Water Research Center, Avondale, PA 19311

Meredith Hullar, Laura Sappelsa, and **David A. Stahl**
Department of Civil and Environmental Engineering, University of Washington, Seattle, WA 98195

Patrick G. Hatcher and **Scott W. Frazier**
Department of Chemistry, The Ohio State University, Columbus, OH 43210

Sponsored by:
Awwa Research Foundation
6666 West Quincy Avenue
Denver, CO 80235-3098

Published by:

DISCLAIMER

Published by IWA Publishing, Alliance House, 12 Caxton Street, London SW1H 0QS, UK

Telephone: +44 (0) 20 7654 5500; Fax: +44 (0) 20 7654 5555; Email: publications@iwap.co.uk;
Web: **www.iwapublishing.com**

AwwaRF report number 91002F. Originally published by AwwaRF for its subscribers in 2003.
IWA Publishing version published 2004.

CONTENTS

TABLES

FIGURES

FOREWORD

The Awwa Research Foundation is a nonprofit corporation that is dedicated to the implementation of a research effort to help utilities respond to regulatory requirements and traditional high-priority concerns of the industry. The research agenda is developed through a process of consultation with subscribers and drinking water professionals. Under the umbrella of a Strategic Research Plan, the Research Advisory Council prioritizes the suggested projects based upon current and future needs, applicability, and past work; the recommendations are forwarded to the Board of Trustees for final selection. The foundation also sponsors research projects through an unsolicited proposal process; the Collaborative Research, Research Applications, and Tailored Collaboration programs; and various joint research efforts with organizations such as the U.S. Environmental Protection Agency, the U.S. Bureau of Reclamation, and the Association of California Water Agencies.

This publication is a result of one of these sponsored studies, and it is hoped that its findings will be applied in communities throughout the world. The following report serves not only as a means of communicating the results of the water industry's centralized research program but also as a tool to enlist the further support of the nonmember utilities and individuals.

Projects are managed closely from their inception to the final report by the foundation's staff and large cadre of volunteers who willingly contribute their time and expertise. The foundation serves a planning and management function and awards contracts to other institutions such as water utilities, universities, and engineering firms. The funding for this research effort comes primarily from the Subscription Program, through which water utilities subscribe to the research program and make an annual payment proportionate to the volume of water they deliver and consultants and manufacturers subscribe based on their annual billings. The program offers a cost-effective and fair method for funding research in the public interest.

A broad spectrum of water supply issues is addressed by the foundation's research agenda: resources, treatment and operations, distribution and storage, water quality and analysis, toxicology, economics, and management. The ultimate purpose of the coordinated effort is to assist water suppliers to provide the highest possible quality of water economically and reliably. The true benefits are realized when the results are implemented at the utility level. The foundation's trustees are pleased to offer this publication as a contribution toward that end.

Natural organic matter (NOM) is important to the quality of drinking water as it constitutes the precursors for disinfectant by-product formation and supports the regrowth of bacteria. The focus of this research is on NOM as a nutrient supply for heterotrophic bacteria and the heterotrophic bacterial community itself. Our work was performed at the interface of microbial ecology and analytical organic chemistry. Little is know about the structure of NOM, what makes certain fractions biodegradable, and how biodegradable organic matter (BOM) influences the composition and dynamics of microbial biofilms? Less, perhaps, is known about the structure, activities, and distribution of the heterotrophic bacterial community. This report presents information that contributes to the basic knowledge of the species composition of microbial communities and the molecular structure of BOM, and begins to reveal some of the complexities as well as commonalities for both subjects. Such information is required by the drinking water industry as it strives to improve biological treatment processes, control the growth of biofilms in distribution systems, and improve BOM measurement techniques.

Edmund G. Archuleta, P.E.
Chair, Board of Trustees
Awwa Research Foundation

James F. Manwaring, P.E.
Executive Director
Awwa Research Foundation

ACKNOWLEDGMENTS

The authors of this report are indebted to the water utilities and academic laboratories for their cooperation and participation in this project. Sherman L. Roberts, Michael D. Gentile, Raphael Morales, and Christian Collado provided technical assistance. The authors thank the Project Advisory Committee members, Dr. Nick Ashbolt, University of New South Wales, Dr. William Anderson, University of Waterloo, and Ms. Ying Wei, City of Houston, for constructive comments that improved this project.

EXECUTIVE SUMMARY

BACKGROUND

Natural organic matter (NOM) is important to the quality of drinking water as it constitutes the precursors for disinfectant by-product formation and supports the regrowth of bacteria. NOM comprises most of the reduced carbon in aquatic ecosystems and provides energy and carbon resources for the metabolism of heterotrophic bacteria. To the extent that water supplies possess different NOM, we believe that the quantity and quality of that resource should be reflected in the community structure of the biofilm growing on the NOM. The drinking water industry is involved in work designed to develop and improve methods for the biological treatment of water, the control of bacterial regrowth in distribution systems, and the measurement of biodegradable organic matter (BOM) concentrations. All of that effort is proceeding without a fundamental knowledge of composition and structure of BOM and NOM, without information on the composition of microbial communities that colonize biological filters and colonize distribution system biofilms, and without knowledge of the dynamics of the microbial populations that compose the microbial communities. Despite these deficiencies, progress is being made, but it is hampered by a lack of basic information.

Our research addressed four major goals: (1) to determine the structure and composition of NOM and BOM, (2) to describe the structure of heterotrophic bacterial communities supported by raw and treated source water, (3) to measure the responses of heterotrophic bacterial communities to seasonally driven variations in NOM and temperature, and (4) to determine whether bioreactor systems can serve as small-scale models for the development and refinement of drinking water treatment processes.

APPROACH

The research project was coordinated through the Stroud Water Research Center under the direction of Dr. Louis Kaplan. The Stroud Center was also responsible for providing assistance with bioreactor construction and the distribution of samples for various analyses. All of the analyses involving molecular microbiology were performed under the direction of Dr. David Stahl of the University of Washington, and all analyses involving nuclear magnetic resonance (NMR) or gas chromatography/mass spectroscopy (GC/MS) were performed under the direction of Dr. Patrick Hatcher of Ohio State University. Participating utilities included Indiana-American Water Company, Inc. in Muncie Indiana, where Dr. Christian J. Volk coordinated the research effort, and the City of Tampa, where Dr. Christine A. Owen coordinated the research effort. The other research participant was the University of Colorado, where work was performed under the direction of Dr. R. Scott Summers.

Five different source waters were selected for this project based on including a broad range of physiographic provinces, vegetation zones (biomes), NOM concentrations, and the abilities and willingness of research participants to install and maintain bioreactors. The study sites included White Clay Creek in the Pennsylvania Piedmont and representative of the Eastern deciduous forest, Hillsborough River in the Florida Coastal Plain and representative of the Southeastern coniferous forest and cypress swamp forest, Barker Lake in the Southern Rocky

Mountains and representative of the alpine/subalpine Western coniferous forest, Rio Tempisquito in the Cordillera de Guanacaste and representative of the tropical evergreen forest, and White River in the Interior Low Plateaus, and representative of the Eastern deciduous forest displaced by agriculture.

We approached the analysis of NOM and microbial communities from a similar analytical hierarchy involving assessment of concentration, composition, and structure. In the case of NOM, concentrations of NOM and BOM were estimated from dissolved organic carbon (DOC) and biodegradable DOC (BDOC) concentrations. DOC was measured with total organic carbon (TOC) analyzers. Composition was assessed from analyses of carbohydrates with ion chromatography with pulsed amperometric detection, humic substances with XAD-8 resin, and functional groups with NMR, and molecular structure was determined from tetramethylammonium hydroxide (TMAH) thermochemolysis GC/MS. For the microbial communities, biomass (analogous to chemical concentration) was determined from direct microscopic counts of bacteria and phospholipid phosphate biomarkers. Community composition was assessed from comparative ribosomal ribonucleic acid (RNA) sequencing, specifically terminal restriction fragment length polymorphisms (t-RFLP), to provide an overview of microbial population structure and detect population shifts at the approximate level of species. A significant amount of the work on this project involved detailed methodological innovations for the t-RFLP and TMAH analyses.

RESULTS

NOM and BOM concentrations showed extensive temporal variation in all of the source waters, but a general pattern of concentration ranges was discernable, indicating that each watershed has a particular concentration signal. Compositional studies revealed that humic substances and complex carbohydrates are components of both NOM and BOM. More detailed structural and compositional studies with TMAH GC/MS identified unique NOM signatures for the different source waters, with some classes of molecules observed only in specific source waters. The combination of TMAH GC/MS analyses with bioreactor processing revealed that the BOM pool includes humic substances and lignin, sources generally presumed to be relatively resistant to biodegradation. Additional novel insights included the quantitative contribution of aromatic molecules to the BOM pool and the potential for bacterial demethylation of lignin. TMAH thermochemolysis GC-MS is demonstrated herein to be effective for investigating the lignin, carbohydrate, and lipid signature of aquatic DOM. Furthermore, the quantitative ability of this procedure adds a new dimension to the comparison of DOM samples within and between laboratories.

The communities of microorganisms that developed in bioreactors fed water from different watersheds were unique. NOM influenced the genetic composition of resulting microbial communities, and seasonal shifts were observed for watersheds possessing strong seasonal temperature signals. Thus, temperature and organic matter quantity and quality probably influenced parameters important to the biological treatment of drinking water. A comparison of bioreactor metabolism with rapid sand filters showed some overlap, suggesting the bioreactors may indicate the ultimate potential of rapid sand filters for BOM processing.

CONCLUSIONS AND RECOMMENDATIONS

The data obtained from TMAH and t-RFLP methodologies support our contention that source waters can possess NOM and BOM of widely different quantity and quality, and that differences in BOM influence the community composition of heterotrophic bacteria that use BOM as a source of carbon and energy. The exploratory nature of our research makes it difficult to prescribe specific treatment steps, but we can recommend the following:

- The species composition of bioreactors and their ability to biodegrade DOC specific to a water source should be recognized when considering bioreactor implementation for monitoring BOM concentrations.
- Bioreactors designed to monitor a BOM source ideally should be inoculated, colonized and maintained by that source; at a minimum, acclimation to the source over several months is needed.
- Bioreactors have potential to aid in water treatment as a means of monitoring BOM concentrations.
- Bioreactors installed and operated in a treatment plant filter gallery should be used to follow the changes in BOM during the treatment process to allow for adjustment in the treatment processes.
- Seasonally driven changes in BOM composition and microbial community composition should be considered when designing biologically active filters.
- Seasonal changes in the microbial community colonizing a biologically active filter may diminish filter performance and require an acclimation period to restore performance.
- Molecular-based methods for both microbial and chemical analyses of drinking water and treatment processes should be targeted for continued development and implementation within the drinking water industry.

CHAPTER 1
INTRODUCTION

BACKGROUND—STATEMENT OF PROBLEM AND SIGNIFICANCE OF THE RESEARCH

Natural organic matter (NOM) is important to the quality of drinking water as it constitutes the precursors for disinfectant by-product formation and supports the regrowth of microorganisms. Our focus was on NOM as a nutrient supply for heterotrophic bacteria and our research is at the interface of microbial ecology and analytical organic chemistry. Bacterial regrowth in a drinking water distribution system leads to the development of biofilms and deterioration of the microbiological quality of the distributed water. NOM is also the target of biological treatment strategies and one of the test variables for measurements of biodegradability. However, little is known about the structure of NOM, what makes certain fractions biodegradable, and how biodegradable organic matter (BOM) influences the composition and dynamics of microbial biofilms.

NOM comprises most of the reduced carbon in aquatic ecosystems and provides energy and carbon resources for the metabolism of heterotrophic bacteria. A central question in organic matter biogeochemistry is what determines whether an organic molecule or compound is biodegradable (Pitter and Chudoba 1990; Sun et al. 1997; Vallino, Hopkinson, and Hobbie 1996)? A major challenge to the study of NOM biodegradability is the limitation of most analytical methods to fully characterize the variable, polymorphous, macromolecular nature of NOM. The characterization of polymeric NOM structure is inherently difficult because, (1) the sheer number of unique compounds present is overwhelming, (2) mass spectrometry (MS), while yielding structural information, requires that analytes be present in the gas phase, but polymeric NOM consists of large, polar and thus nonvolatile molecules, and (3) the complexity of mass- and nuclear magnetic resonance (NMR) spectra increases with molecular size and complexity of the analyte.

A central question in microbial ecology is how cosmopolitan are the various species of microbiota? To the extent that water supplies possess different NOM, the quantity and quality of that resource should be reflected in the community structure of the biofilm growing on the NOM. However, the difficulty of identifying species of bacteria, including but not limited to the selective nature of cultivation, means that until most recently, little has been known about the species composition and variability of natural bacterial communities (Lee and Fuhrman 1990). Advances in molecular techniques have made possible the phylogenetic identification and detection of microbial cells without cultivation (Amann, Ludwig, and Schleifer 1995; Stahl 1995) and has extended identifications within complex communities to the species or even subspecies levels (Amann, Ludwig, and Schleifer 1994; Raskin, Rittmann, and Stahl 1996).

Studies from the marine environment, including samples from the Pacific and Atlantic Oceans (Schmidt, Delong, and Pace 1991; Fuhrman, Mccallum, and Davis 1993) have found strong similarities between microbial communities in these disparate environments, but that has not been the case for soils. Soil samples collected from four Mediterranean and two boreal ecosystems at sites selected with similar climates, soils, and functional vegetation type had communities composed of genetically distinct strains (Fulthorpe, Rhodes, and Tiedje 1996; Zhou et al. 2002). To the extent that organic matter resources influence the structure of bacterial communities within aquatic and terrestrial environments, these data suggest that marine phytoplankton degradation products (Aluwihare and Repeta 1999) and exudates are similar, but that biological, chemical, and physical changes of soil organic matter (diagenesis) produces highly specific end products. Additionally, physical mixing of the species pool should play an important role in the relative genetic heterogeneity of bacterial communities within and between habitats. Ocean mixing, albeit slow, far exceeds the mixing of soils that can be accomplished through the activities of soil invertebrates (bioturbation) or aerial dispersal. We suggest that a combination of organic matter heterogeneity and physical mixing generate the differences in bacterial community composition reported for these habitats, and that streams, rivers, and even ground waters are reasonably similar to the soil condition in organic matter heterogeneity and species diversity.

If we extend this reasoning to drinking water sources, it follows that most source waters should have a watershed NOM signal that results in the selection of characteristic, if not unique assemblages of microorganisms. Those organisms constitute the species pool that is the source of organisms that colonize biological treatment filters and form distribution system biofilms. Another indication that watershed NOM is highly unique comes from the homing behavior of migratory fishes. Stream and river water NOM provides an olfactory imprint of adequate specificity and persistence to guide fishes to the particular stream or river where they were born (Hasler and Wisby 1951; Scholz et al. 1976; Dittman, Quinn, and Nevitt 1996). The NOM of terrestrial origin present in river ecosystems is a product of soil processes and the watershed specificity and persistence of that NOM signal could select for unique assemblages of bacterial heterotrophs at treatment plant influents. We do not have the information currently needed to confirm this supposition, but our experience with biofilm reactors suggests it is correct.

Biofilm reactors colonized and supported by streamwater have been used to measure concentrations of BOM in drinking waters (Lucena, Frias, and Ribas 1990; Kaplan and Newbold 1995; Anderson, Urfer, and Huck 1997; Prevost et al. 1997) and to characterize the composition of the BOM pool (Volk, Volk, and Kaplan 1997). The unidirectional flow of nutrients and energy in the bioreactors make these model systems aquatic analogs of the Winogradsky soil column, and generate gradients of NOM concentration, NOM composition, microbial biomass, activity, and community composition (Kaplan, Bott, and Frias 1994). We isolated 282 bacteria from the bioreactors and grew them in pure culture, and prepared samples for fatty acid methyl ester analyses using the MIDI system. Many of the colonies gave "no match" with the identification library, but we were able to determine that there were 120 unique isolates and that the distribution of isolates throughout the bioreactors was variable (Kaplan, Bott, and Frias 1994).

In characterizing the composition of the BOM pool with bioreactors, we were surprised that 75% of the BOM in streamwater was contributed by humic substances (Volk, Volk, and Kaplan 1997), and even 54% of the most biologically labile BOM measured was humic substances (Kaplan and Gremm 1995 a,b). Additional support for the observation that polymeric NOM is an important component of BOM comes from kinetic studies of a drinking water distribution system biofilm supported by humic substances as the primary carbon source (Butterfield et al. 1997). An equally intriguing observation of bioreactor performance was that bioreactors colonized on and acclimated to NOM in one watershed exhibited a high degree of metabolic specificity for that NOM. Bioreactors colonized by water from White Clay Creek in the eastern deciduous forest of Pennsylvania and other bioreactors colonized on water from the River Oise in France were unable to metabolize BOM from Bull Run in the coniferous forest of the Oregon Cascades, while bioreactors colonized on Bull Run water routinely metabolized 100 to 200 µg C/L of the Bull Run NOM (Kaplan et al. 1996). These experiments involved short-term exposures (<24h) and included the addition of mineral salts to insure carbon limitation.

2

In a long-term experiment, 25 bioreactors were colonized on water from the Mississippi River drainage, collected near St. Louis. Once colonized and functional, these bioreactors were installed at different watersheds throughout North America. Those installed at sites within the Mississippi drainage worked well (metabolized BOM) while the others gave mixed results, but generally did not perform well. Following 8 months of operation however, all bioreactors operated well, with the change over requiring approximately 5 to 7 months of continuous exposure to site waters and the indigenous microorganisms (Volk, Volk, and Kaplan 1997; Kaplan and Volk 1997). These data augment our observations of metabolic specificity within a watershed. The time frame of the response suggests that there was a replacement of species within the bioreactors by species indigenous to the new watersheds rather than an enzymatic adjustment of the existing species or the selective growth of populations within the original community.

From the perspective of a drinking water treatment plant operator, imagine trying to control the activity of microorganisms (e.g. limit growth and colonization within distribution systems or increase NOM biodegradation to CO_2 within treatment filters) without a good understanding of the food resources for those organisms, the composition of the microbial community, or the natural history of the populations within that community? This is, in fact, the situation that drinking water utilities face. The drinking water industry is involved in work designed to develop and improve methods for the biological treatment of water, the control of bacterial regrowth in distribution systems, and the measurement of BOM concentrations. All of that effort is proceeding without a fundamental knowledge of composition and structure of BOM and NOM, without information on the composition of microbial communities that colonize biological filters and colonize distribution system biofilms, and without knowledge of the dynamics of the microbial populations that compose the microbial communities. Despite these deficiencies, progress is being made, but it is hampered by a lack of basic information.

Our work addressing NOM characterization and microbial community structure, casts a wide net around two exceptionally complex issues. We have no illusions that we have completely answered questions that are at the forefront of the fields of organic matter biogeochemistry, microbial ecology, and theoretical ecology. Nevertheless, we are confident that we have substantively improved the knowledge base that drinking water professionals can draw upon to address the treatment of NOM and the regrowth of microorganisms. For example, the information generated by this study has direct application to attempts to improve NOM removal through rapid filtration and improve BOM measurement techniques with bioreactors. Additionally, the methods that we have applied to these questions, often for the first time within the drinking water industry, will help future research projects better understand the strengths and weakness of these analytical approaches. Finally, we suggest that an understanding of BOM chemistry and information about the structure of microbial communities supported by BOM are an essential knowledge base that is required for the drinking water industry as it strives to improve (1) biological treatment processes, (2) the control of biofilm growth in distribution systems, and (3) BOM measurement techniques.

RESEARCH GOALS AND OBJECTIVES

Our study had four major goals, each of which included several specific research objectives. Below we list the goals and the underlying objectives that guided our research efforts:

1. **Determine the structure and composition of NOM and BOM.**
   a. Characterize the NOM and BOM pools from different source waters.

b. Characterize the seasonal changes in the quantity and quality of NOM and BOM.

c. Assess the influence of treatment on NOM and BOM.

2. Describe the structure of heterotrophic bacterial communities supported by raw and treated source water.

a. Compare the species composition of heterotrophic bacterial communities in source waters from different watersheds.

b. Test the ability of a community of heterotrophic bacteria from one watershed to metabolize the NOM generated within a different watershed.

3. Measure the responses of heterotrophic bacterial communities to seasonal changes driven by variations in NOM and temperature.

a. Assess the response of a bacterial community to the combined effects of seasonal changes in NOM and temperature.

b. Assess the response of a bacterial community structure to the effect of seasonal changes in NOM.

c. Assess the response of a bacterial community structure to the effect of seasonal changes in temperature.

4. Determine whether bioreactor systems can serve as small-scale models for the development and refinement of drinking water treatment processes.

a. Compare the microbial communities in bioreactors to those in the natural communities.

b. Compare the microbial communities in bioreactors to those in treatment filters.

c. Assess the ability of bioreactors to predict the performance of biological filters used in water treatment.

CHAPTER 2
EXPERIMENTAL APPROACH

PROJECT COORDINATON AND RESEARCH PARTICIPANTS

The research project was coordinated through the Stroud Water Research Center under the direction of Dr. Louis Kaplan. The Stroud Center was also responsible for providing assistance with bioreactor construction and the distribution of samples for various analyses. All of the analyses involving molecular microbiology were performed under the direction of Dr. David Stahl of the University of Washington, and all analyses involving NMR or GC/MS were performed under the direction of Dr. Patrick Hatcher of Ohio State University. Participating utilities included Indiana-American Water Company, Inc. in Muncie Indiana, where Dr. Christian J. Volk coordinated the research effort, and the City of Tampa, where Dr. Christine A. Owen coordinated the research effort. The other research participant was the University of Colorado, where work was performed under the direction of Dr. R. Scott Summers.

SELECTION OF STUDY SITES

Five different source waters were selected for this project based on including a broad range of physiographic provinces, vegetation zones (biomes), NOM concentrations (assessed from concentrations of dissolved organic carbon (DOC)), and the abilities and willingness of research participants to install and maintain bioreactors (Table 2.1).

BIOREACTOR AND RAPID SAND FILTER COLONIZATION

Bioreactors constructed and operated following the design and procedures of Kaplan et al. (1996) were established with source water and the associated microorganisms from each of the sites. The bioreactors were used to identify the BOM component of dissolved organic matter (DOM) through the measurement of DOC in the bioreactor influent and effluent waters. The difference between DOC concentrations in the inflow and outflow is considered the biodegradable DOC, while the organic matter remaining in the effluent is considered refractory DOM (RDOM). All feed waters, with the exception of those from the White River, discussed below, were filtered through glass fiber filter cartridges (75 μm, 25 μm and 3 μm in series, (Balston, Tewksbury, MA)) that remove large particles but allow suspended bacteria to pass through (Kaplan and Newbold 1995). At White Clay Creek and Rio Tempisquito, seasonal impacts on bioreactor colonization and operation were assessed with water-jacketed bioreactors. The water jackets were fed with stream water to maintain the bioreactors at stream water temperatures during the colonization and operation period. At White Clay Creek, the treatments involved bioreactors from cold weather (October through April) and warm weather (May through November) periods, while at the Rio Tempisquito, where annual water temperatures vary by 1°C (Newbold et al. 1995), the bioreactors represented wet-season (May through November) and dry-season (November through May) conditions.

The two drinking water utilities involved in bioreactor operation were Indiana American in Muncie and City of Tampa in Florida. At Tampa, the bioreactors were supplied with the Balston-filtered water from the Hillsborough River. In a test of a continuous filtration system for bioreactors operated on raw water, White River water at the Muncie water treatment plant was filtered

Table 2.1
Characteristics of source waters

Water source (temperature range)	Physiographic province	Vegetation zone	DOC range (mg C/L)	Municipality served
White Clay Creek (0° to 24°C)	Pennsylvania Piedmont	Eastern Deciduous Forest	0.7 to 12	Wilmington, Delaware
Hillsborough River (13° to 29°C)	Florida Coastal Plain	S. Eastern Coniferous Forest/Cypress Swamp	3 to >30	Tampa, Florida
Barker Lake (0° to 22°C)	Southern Rocky Mountains	Alpine/Subalpine Western Coniferous Forest	2 to 6	Boulder, Colorado
Rio Tempisquito (20° to 22°C)	Cordillera de Guanacaste	Tropical Evergreen Forest	0.4 to 4.5	Maritza Biological Station
White River (2.5° to 26.5°C)	Interior Low Plateaus	Eastern Deciduous Forest/Agriculture	2 to 6	Muncie, Indiana

through a 3-5 μm stainless steel membrane using an on-line filtering system (Model 9700, Collins Products Co., Livingston, TX). The filtered water was collected in a 120-L reservoir and continuously fed through duplicate bioreactors. Colonization of the bioreactors on White River water began in March 1999. Bioreactors fed by water from the Hillsborough River near Tampa were begun in November 1999 and involved 4 replicate bioreactors.

At the University of Colorado, bioreactors and rapid sand filters (RSF) were established at the City of Boulder's Betasso water treatment plant fed by Barker Lake water. A total of eight systems were established, four rapid sand filters and four bioreactors with SIRAN beads (Jaeger Biotech Engineering, Inc., Costa Mesa, CA) as the medium. The four sets of RSF were started in February 1999 and DOC sampling began at the end of that month. The RSF utilized quartz filter sand (effective size 0.44 mm) and were operated at 35 mL/min. The filter consisted of three chromatographic columns (600mm length and 25 mm diameter; hydraulic loading of 4.28 m/h) in series in which the first column always represents a three-minute empty-bed contact time (EBCT) and the third a seven-minute EBCT. The experimental design for the RSF system was developed to assess the effect of treatment process (ozonation) and temperature (Figure 2.1). The system testing intermediate ozonation consisted of two columns with an EBCT of three minutes followed by a third one with 7 minutes EBCT. Three RSF were run at ambient temperature (15°C to 22°C): one fed raw water; a second fed by pre-ozonated raw water (average transferred O_3/DOC of 1.0 mg/mg); and a third fed by water from an intermediate ozonation step (average transferred O_3/DOC of 1.0 mg/mg), after the first column. Running a RSF fed by pre-ozonated water at 3°C rather than at ambient temperatures assessed the effect of temperature. The bioreactors at Betasso water treatment plant were begun in May 1999 and included raw water and pre-ozonated water as the feeds.

BIOREACTOR CHALLENGES

Bioreactors from the Betasso water treatment plant and from the City of Tampa were subjected to cross-feeding experiments. The experiments involved shipping the bioreactors that had been colonized and operated on Barker Lake and Hillsborough River waters, respectively, to the Stroud Center and feeding them with water from White Clay Creek under two treatments. One

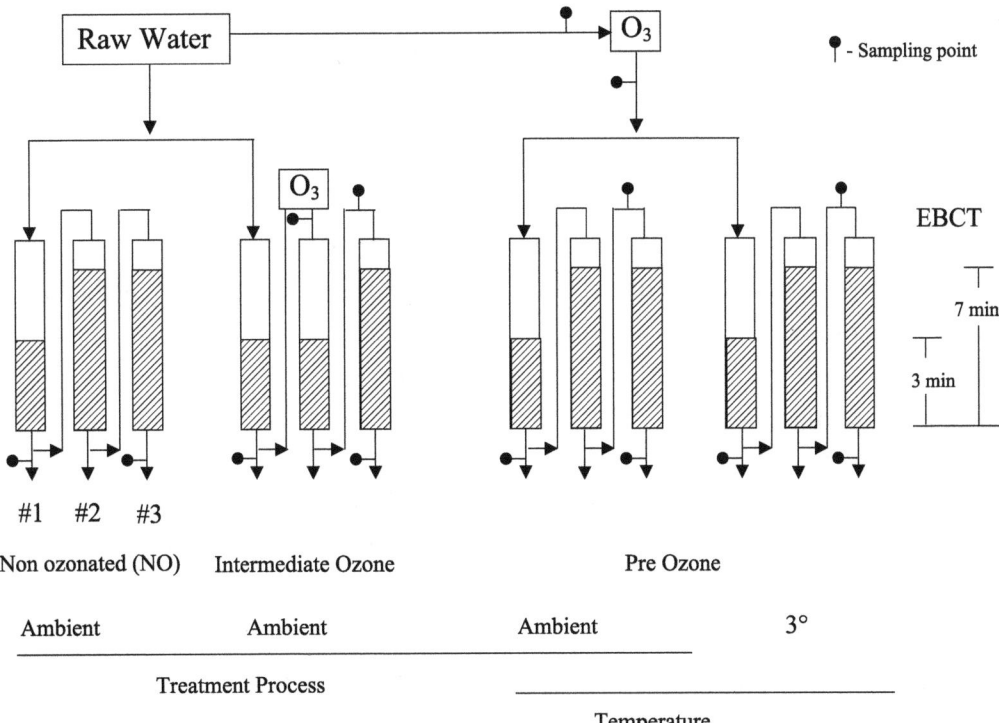

Source: Adapted from *Water Research,* Vol. 35, No. 16, C. Fonseca, R.S. Summers, and M.T. Hernandez, Comparative Measurements of Microbial Activity in Drinking Water Biofilters, pp. 3817-3824, Copyright 2001, with kind permission from Elsevier Science, Ltd., The Boulevard, Langford Lane, Kidlington OX5 1GB, UK.

Figure 2.1 Betasso water treatment plant biological filter set-up with empty-bed contact times (EBCT)

treatment involved using Balston filtered White Clay Creek water with its full compliment of bacteria and the other involved using sterile White Clay Creek water. Initially, White Clay Creek water was sterilized by autoclaving. The autoclaved White Clay Creek water was used in a 7-month long experiment with bioreactors from the Betasso water treatment plant (November 1999 – June 2000). Eventually, for the experiment with the Hillsborough River bioreactors, the stream water was filter sterilized with 0.1 μm Gelman PALL cartridges into autoclaved carboys. The Hillsborough bioreactors were challenged for a period of 1 month (November 2000 – December 2000). A single 20-L carboy was used with 1 bioreactor, so the filter-sterilized water was held over a period of approximately 3 days at temperatures ranging from 15°C to 20°C. Sterile cotton plugs in the vent lines and sterilized Pharmed® tubing on the supply lines, operated with quick disconnect fittings were used to ensure sterility. Prior to the challenge, replicates of the test bioreactors were destructively sampled to characterize the bacterial community, and at the end of the challenge, the sampling was repeated.

Table 2.2
Analytical methods

Parameter	Reference
Microbial Community	
Biomass	
Phospholipid phosphate	Findlay, King, and Watling 1989
Direct microscopic counts total cells	Bott and Kaplan 1993; Hobbie, Daley, and Jasper 1977
Community Composition	
T-RFLP	Liu et al. 1997
Natural Organic Matter/Biodegradable Organic Matter Concentration	
DOC	Kaplan 1992
BDOC	Kaplan and Newbold 1995; Volk, Volk, and Kaplan 1997
Composition	
Carbohydrates	Gremm and Kaplan 1997
XAD-8 resin chromatography	Thurman and Malcolm 1981
^{13}C-NMR	Hatcher, VanderHart, and Earl 1980
Structure	
TMAH/GC/MS	del Rio and Hatcher 1996

ANALYTICAL PROTOCOL OVERVIEW

We approached the analysis for NOM and microbial communities from a similar analytical hierarchy involving concentration, composition, and structure (Table 2.2). In the case of NOM, concentrations of NOM and BOM were estimated from DOC and BDOC concentrations with TOC analyzers, composition was assessed from analyses of carbohydrates with ion chromatography with pulsed amperometric detection, humic substances with XAD-8 resin, and functional groups with NMR, and molecular structure was determined from tetramethylammonium hydroxide (TMAH) thermochemolysis GC/MS. For the microbial communities, biomass (analogous to chemical concentration) was determined from direct microscopic counts of bacteria and phospholipid phosphate, and community composition was assessed from comparative ribosomal RNA sequencing, specifically terminal restriction fragment length polymorphisms (t-RFLP), to provide an overview of microbial population structure and detect population shifts at the approximate level of species. A significant amount of the work on this project involved detailed methodological innovations for the t-RFLP and TMAH analyses, so we have described those methods in considerable detail below.

MICROBIOLOGICAL ANALYSES

Epifluorescence Microscopic Counts

Bacterial abundance was measured for sediment, bioreactor, and rsf samples by epifluorescence microscopic counts (EMC) following staining with propidium iodide (Bott and Kaplan 1993). Samples were preserved in formalin, sonicated for 1.5 min. at 50 to 70 W in 0.1 M dibasic ammonium phosphate to detach the bacterial cells from the substratum (Velji and Albright 1986). Following sonication, 2 mL of 60% glycerol were added and the samples were vortexed and centrifuged for 3 min at 853 × gravity to separate the bacteria from mineral particles. Bacteria were stained with 5 drops of 0.2 mg/L propidium iodide and filtered onto a black 0.2 μm polycarbonate filter. Bacterial cells were enumerated in 20 fields with a Zeiss Universal microscope.

Phospholipid Phosphate

Bacterial biomass was estimated for bioreactor and rapid sand filter samples by the determination of phospholipid phosphate. Lipids were recovered from all samples by chloroform - methoanol extraction, phosphate liberated from the lipids by potassium persulfate digestion, and the phosphate released by digestion was determined by reaction with ammonium molybdate and spectophotometric detection (Findlay, King, and Watling 1989).

DNA Extraction

The use of DNA-based methods to characterize samples recovered from heterogeneous environments, such as bioreactors, rapid sand filters, or streams, generally requires optimization of existing molecular protocols to minimize inhibition of enzymes by compounds that naturally co-occur with DNA extracted from different matrix materials. Methods were optimized at several levels of data gathering and analysis. DNA extraction was optimized to remove contaminating compounds such as humic material. After extraction of DNA from samples using standard phenol: chloroform:isoamyl alcohol procedures, several procedures were tested to remove humic material in order to produce amplifiable DNA. Two gel based approaches, using either polyvinylpyrrolidone (PVP) gels (Young et al. 1993) or low melt agarose gels (FMC, Philadelphia PA), failed to produce DNA that was pure enough for polymerase chain reaction (PCR) amplification. An alternative extraction procedure, using a phosphate buffer:phenol and bead beating disruption protocol modified from the Fast DNA kit (QBiogen, Carlsbad, CA) produced good quality genomic DNA that was amplifiable using the PCR.

DNA was extracted from all samples using a "bead beater" protocol. Approximately 1 gram of each sample was collected in a 2-mL microcentrifuge tube containing 0.5 g of zirconium beads (0.1 mm diameter, Biospec Barltesville, OK -baked at 200°C for at least 2 hours). Phosphate and MT Buffers from the Fast DNA kit were added to the sample tubes. The samples were processed in a Bio101 FastPrep (QBiogene, Inc. Carlsbad CA) at speed 4.5 for 15 seconds and then placed on ice for 1 minute. They were then processed in the FastPrep at speed 4.5 for another 15 seconds. The samples were centrifuged at 4°C at 15000 rpm for 5 minutes. For each sample, the supernatant was placed into new tubes and samples were processed according to the kit protocol. Samples were eluted in 50 μL diethyl pyrocarbamate (DEPC)-treated water and stored at -80°C. To check quality, the DNA was loaded on a 0.8% (w/v) high-melt agarose gel with 1× running buffer (40 mM Tris-acetate and 1mM EDTA, pH 8.3) and electorphoresed at 100 V/cm for 35 minutes. DNA was visualized using ethidium bromide (Sambrook, Fritsch, and Maniatus 1989). This protocol varied from standard DNA extraction protocols by using phosphate buffers and bead beating with zirconium beads to extract DNA from cells.

Polymerase Chain Reaction (PCR)

PCR was optimized for the different environmental systems using a combination of touchdown PCR and the addition of bovine serum albumin (BSA; Sigma) to the PCR reactions to bind environmental contaminants that otherwise inhibited PCR. Touchdown PCR involves decreasing the annealing temperature by 1°C every second cycle to a "touchdown" annealing temperature, which is then used for approximately 10 cycles (Don et al. 1991). Because any difference in temperature between a correct and an incorrect annealing gives a 2-fold difference in product amount per cycle (4-fold per °C), this procedure helps to enrich for the correct product over any incorrect products. We found that higher levels of protein (4%) than commonly added to PCR reactions were needed to achieve amplification. Genomic DNA samples were diluted 1/100 in DEPC water. PCR was performed using bacterial primers S-D-Bact-008-a-S-27 (Giovannoni et al. 1988) that was 5'end-labeled with 6-carboxyfluroescein (6-FAM) and S-D-Bact-1512-a-A-21 (Giovannoni et al. 1988). Each 50 μL PCR reaction contained both primers (0.4 M), 10 mM Tris-HCl (pH8.8), 50 mM KCl and 1.5 mM $MgCl_2$, 200 M of each deoxynucleoside triphosphate (Idaho Technologies, Salt Lake City, Utah), 25 μg of 10× BSA and 2.5 units (U) Taq DNA polymerase (Amersham, Carlsbad, CA). Samples were amplified on a Hybaid PCR Express thermal cycler (Hybaid, UK) using the following "touchdown" parameters: initial denaturation step of 94°C for 1 minute, followed by 4 cycles of denaturation at 94°C for 30 seconds, annealing at 65°C for 30 seconds followed by extension at 72°C for 30 seconds, annealing at 65°C for 30 seconds plus temperature decrease at 1°C per cycle and extension at 72°C for 30 seconds, then 5 cycles of denaturation at 94°C for 30 seconds, annealing at 50°C for 30 seconds and extension at 72°C for 30 seconds. Cycling was completed by a final extension at 72°C for 5 minutes. The PCR products were visualized on a 0.8% (w/v) high-melt agarose gel. A combination of dilution of genomic DNA, higher levels of protein in the PCR reaction mix, and touchdown PCR was required to achieve amplification of environmental DNA.

Terminal Restriction Fragment Length Polymorphism (t-RFLP) Analysis

t-RFLP is an analytical method that differentiates components of the bacterial community based on sequence differences of the 16S rRNA gene (Liu et al. 1997). We standardized the t-RFLP analysis to increase reproducibility of t-RFLP traces via the development and optimization of software (DAx, van Mierlo Inc., Eindhoven, The Netherlands). The software program uses linear regression to identify the migration of each fragment using an internal lane standard. Each sample was digested using PCR product, 10U of a tetrameric restriction enzyme (HAE III), and restriction buffer (One-Phor-All Plus, Amersham-Pharmacia). Samples were incubated in the dark at 37°C (HAE III) or 60°C for at least 4 hours. Digested samples were ethanol precipitated, dried and sent to Oregon State University Center for Genomic Research for analysis. At least 100 fmol of digested sample was loaded onto a 4% polyacrylamide gel for fragment analysis using an ABI 377 DNA sequencer (Amersham, Piscataway, NJ) in Gene Scan mode. Genescan 500 TAMRA® size standard (Amersham, Piscataway, NJ), a fluorescent dye conjugated to size standards was

10

added to each sample lane on the gel. ASCII files of electropherograms were analyzed using DAx analysis software.

Gene Cloning, Sequencing and Phylogenetic Inference

Amplified products were ligated into the pCR4 cloning vector and transformed into ONE SHOT competent *Escherichia coli* cells following the manufacturer's instructions (TOPO TA Cloning Kit for Sequencing, Invitrogen, Carlsbad, CA). Plasmid DNA was isolated using the Wizard MiniPrep kit (Promega). Individual clones were sequenced on a Megabace capillary system at the University of Washington's Center for Marine Biotechnology. All sequences were aligned and phylogenetic trees constructed using the neighbor-joining algorithms implemented by Phylogeny Inference Package (PHYLIP) in the arbor (ARB) software program (www.arb-home.de). Unique t-RFLP fragments were identified using Fragment Finder, software we developed to facilitate the *in-silico* analysis of DNA sequences from clone libraries. Fragment Finder is available at http://stahl.ce.washington.edu/. All phylogenetic trees were constructed using a mask of 608 informative bases from positions 1-667 (*E. coli* numbering). Twenty-three unique clones were analyzed from samples collected in May 1999 from a Rio Tempisquito bioreactor, streambed sediments, and organisms that grow on rock surfaces in a riffle (i.e., epilithon).

CHEMICAL ANALYSES

Dissolved Organic Carbon

Dissolved organic carbon (DOC) was measured by Pt-catalyzed persulfate oxidation using either an OI 700 or an OI 1010 carbon analyzer (Kaplan 1992). All samples except for bioreactor effluents were filtered through precombusted (500°C/6h) glass fiber filters (Whatman GF/F) prior to analysis. The bioreactors have a 10 μm pore-sized support disk that effectively removes particles from the bioreactor effluent. Humic substances were measured by macroreticular XAD-8 resin chromatography (Thurman and Malcolm 1981). Humic substances were defined as the difference between the DOC concentration of samples measured prior to acidification and the effluent from the XAD-8 resin.

Carbohydrates

Dissolved total carbohydrates (DTCHO) were determined by high-performance liquid chromatography with pulsed amperometric detection (HPLC-PAD). Analyses were performed on a Dionex 500 with a PA-1 column (Gremm and Kaplan 1997). All samples were prefiltered through 0.2 μm pore-sized membrane filters.

Cross-Polarization Magic Angle Spinning ^{13}C Nuclear Magnetic Resonance (CPMAS ^{13}C NMR)

High field ramp cross-polarization magic angle spinning (CPMAS) ^{13}C nuclear magnetic resonance (NMR) experiments were performed using high speed spinning with a ramp CPMAS pulse program and two pulse phase modulation (TPPM) decoupling. The ramp CPMAS pulse program was employed as described by Metz, Wu, and Smith (1994) with the carbon spin-locked power ramped linearly from one half its final value. Background cross-polarization (CP) spectra were acquired and found to be insignificant. Traditional CPMAS experiments were performed with constant amplitude spin lock fields that were reoptimized for maximum polarization transfer. Hartmann-Hahn matching and TPPM parameters were optimized on polycrystalline glycine under similar conditions as those employed on the humic acids. The carbonyl shift of glycine (176.03 ppm) provided a convenient secondary reference for all the solid-state NMR spectra.

The contact time dependence of the CP was measured by performing a series of experiments in which the contact time was lengthened from 20 µs to 15 ms. As will be demonstrated for our samples, a contact time of 1 to 2 ms was found to give close to quantitative results, when a recycle delay time of 1 s and a sample spinning speed of 13 kHz were used. In all presented spectra, decoupling field strengths were greater than 70 kHz and spin locking field strengths were greater than 50 kHz. Free induction decays over a sweep width of 27778 Hz with 1024 complex data points were collected and zero filled to 8192 total complex data points. A 100 Hz exponential line broadening was applied before Fourier transformation and phasing.

Tetramethylammonium Hydroxide (TMAH) Thermochemolysis

We selected a degraded lignin sample to quantitatively evaluate the reproducibility of the TMAH thermochemolysis reaction. The degraded lignin sample was from a Douglas Fir (*Pseudotsuga menziesii*) collected from Mt. Rainier, WA and was extensively described in a previous communication (Hatcher et al. 1995). The wood was well degraded so that, presumably, mostly lignin remained with little cellulose (Hatcher 1987).

Water samples from streams or bioreactor effluents were collected in polycarbonate containers, frozen, and transported back to the lab where they were thawed and then concentrated using low-temperature, low-pressure evaporation (B.Ü.C.H.I., Flawil, Switzerland) and then freeze-dried (Labconco, Kansas City, MO) to yield the total dissolved solids (TDS). The dried samples were between 0.5 and 3% C by weight as they contained mostly salts. In previous experiments salts were found to have minor effects on TMAH product distribution (Milano, personal communication).

Elemental analyses were performed primarily for normalizing the quantitative TMAH thermochemolysis GC-MS results. Dr. Matt Charvette at the Woods Hole Oceanographic Institution (Woods Hole, MA) provided the elemental analysis data for the isolated dissolved material. To remove inorganic carbon the solid sample was vapor acidified with HCL for 24 hours and then baked at 50°C for one hour to remove residual HCL and water prior to analysis.

TMAH Thermochemolysis Procedure for Mt. Rainier Lignin

For the TMAH thermochemolysis analysis of the Mount Rainier degraded lignin, the sample was ground with a mortar and pestle and 1.0 mg of the sample was placed in a glass ampoule with 200 µL of TMAH (25% in methanol). The methanol was evaporated under a stream of nitrogen gas and the glass ampoule was sealed under vacuum and then baked at 250°C for 30 minutes. After the ampoules were cooled they were scored, frozen in liquid nitrogen to prevent loss of products, and broken open at which time 50 µL of a 38 ng/µL internal standard (eicosane) solution in ethyl acetate (1.9 µg total internal standard) were added. The products were extracted from the ampoule with 1 mL of ethyl acetate and concentrated in a GC vial under a stream of nitrogen gas to approximately 200 µL. These samples were prepared in triplicate.

TMAH Thermochemolysis Procedure for Dissolved Organic Matter (DOM)

For the TMAH analysis of the DOM samples, approximately 100 mg of the isolated total dissolved solids (TDS) (0.5-3 mg of organic C) was weighed out into two separate glass ampoules (50 mg in each) and 200 μL of TMAH (25% in methanol) was placed in each ampoule. We found that using two ampoules as opposed to one for this much sample minimized loss of sample to "explosions." The ampoules were completely mixed to saturate the entire sample and the methanol was evaporated under a stream of nitrogen gas. The glass ampoules were sealed under vacuum and baked at 250°C for 30 minutes. After the ampoules were cooled they were scored, frozen in liquid nitrogen, and broken open at which time 20 μL (40 μL total per sample) of a 9.40 ppm internal standard (eicosane) solution in ethyl acetate were added to each ampoule (376 ng total). The products were extracted from the ampoule with 2 mL of ethyl acetate and concentrated under nitrogen gas to approximately 100 μL in a GC vial prior to analysis by GC-MS.

Optimizing Internal Standard Addition

The quantitative GC-MS analysis of the TMAH products uses an internal standard approach, which is discussed in detail below. Therefore, an experiment was performed to determine the optimal point during the TMAH thermochemolysis procedure to add the internal standard. To obtain the most reproducible and representative results, the internal standard should be added to the sample as early in the procedure as possible. However, we were concerned that the TMAH reaction would degrade the internal standard if it were added to the sample prior to the TMAH reaction.

We designed this experiment to determine if standards are degraded during the reaction and to determine if there are differences in the recoveries of different standards (and products) during the TMAH product extraction procedure. To assess these, we added a mixture of standard compounds consisting of TMAH products commonly observed from lignin (standards described in the quantitative TMAH section below), hexyl benzene and eicosane (C_{20} alkane) at different points during the TMAH procedure. We chose to use eicosane as the internal standard for the quantitative analysis of the samples in this study since previous results indicated this compound would not overlap with products produced from the TMAH thermochemolysis reaction with lignin and DOM (Zang et al. 2000; Fabbri and Helleur 1999).

For this experiment we selected a DOM sample collected from the Schuylkill River as part of a separate study. Schuylkill River water was collected from the raw water intake of the Norristown Water Treatment Plant of the Pennsylvania American Water Company. The Schuylkill has both industrial and wastewater influences and the DOC ranges from 2-6 mg L^{-1}. The DOM sample (50 mg) was added to the TMAH ampoules along with 200 μL of TMAH (25% in methanol). A lignin standard solution (containing the standards described below) containing eicosane was prepared in ethyl acetate to a final concentration of 15-25 ng μL^{-1} for each of the standards. The DOM samples were spiked with 20 μL of the lignin standard solution containing eicosane (approximately 2-3 ng of each of the standards) either prior to sealing the ampoule or prior to extraction. Two sets of three ampoules were prepared. To the first set, the lignin and eicosane standard solutions were added to the ampoules prior to the addition of TMAH, evaporating the methanol, sealing under vacuum, baking, and extraction. To the second set, the lignin and eicosane standard solutions were added to the ampoules after the TMAH reaction, but before the product extraction described above.

Another set of ampoules was prepared with this DOM sample (50 mg) though only eicosane was added to this set after the TMAH reaction. This was done to determine the yield of the sample-derived lignin compounds that overlap with those in the standard solution. Corrections were made for the sample-derived, lignin products so that recoveries could be determined relative to eicosane. The lignin-standard solution was also analyzed separately using GC-MS (without addition to DOM). The GC-MS results for the lignin-standard solution were compared with those of the DOM samples to which the standard additions were made.

Instrumental Aspects

The TMAH products were analyzed on a Hewlett-Packard 6890 GC with a split-splitless injector operating in the split mode and using helium as the carrier gas. The GC was fitted with a 15 meter Rtx-5MS® capillary column (0.25 mm i.d., 0.1 μm film thickness, Restek, Bellefonte, PA). The GC was coupled to a Pegasus II® (Leco Corporation, St. Joseph, MI) time-of-flight mass spectrometer operating in the electron impact (EI) mode with a filament bias of –50 V. The source was maintained at a temperature of 200°C and the transfer line at 280°C. Peak identifications were based upon comparison with standards and the National Institute of Standards and Technology (NIST) (version 1.6) library.

Separate GC programs were used to obtain either qualitative or quantitative results. To facilitate qualitative comparisons and presentation, the GC had a constant inlet temperature of 310°C and the temperature was programmed from 40°C to 310°C at a rate of 8°C min^{-1} holding for 1 minute at the minimum and maximum temperatures. Mass spectra were collected at a rate of 15 spectra per second after the 120-second solvent delay. Masses were acquired between m/z 35 and 500. Due to the nature of this GC-MS system it is possible to obtain more precise and accurate results using faster GC programs. Co-eluting compounds can be completely resolved from one another by using unique m/z or the proprietary algorithm that deconvolutes the areas of co-elutions using a chemometric approach. Therefore, for the quantitative analysis of DOM samples the GC had a constant inlet temperature of 310°C and the temperature was programmed from 50°C to 300°C at a rate of 25°C min^{-1} holding for 1 minute at the minimum and maximum temperatures. Mass spectra were collected at a rate of 35 spectra per second after a 115 sec. solvent delay. Masses were acquired between m/z 35 and 400.

Quantitative Analysis of Products of TMAH Thermochemolysis on the GC-MS

Previous studies (Zang et al. 2000; Fabbri and Helleur 1999) indicated that the TMAH treatment of DOM produces a suite of products that include fatty acid methyl esters (FAMEs), lignin-derived aromatic products, as well as aromatic products from other sources. Standards were prepared for FAMEs with even number carbon chains ranging from octanoic acid methyl ester through tetracosanoic acid methyl ester with selected unsaturated FAMEs. A set of standards for the more common lignin-derived TMAH products resulting from DOM has been compiled for: (1) p-hydroxyphenyl compounds: 1-ethenyl-4-metahoxy-benzene; 4-methoxy-benzaldehyde; 1-(4-methoxy phenyl) ethanone; 4-methoxy-benzoic acid, methyl ester; (2) guaiacyl compounds: 1,2-dimethoxy benzene; 3,4,dimethoxytoluene; 3,4-dimethoxy-benzaldehyde; 1-(3,4-dimethoxyphenyl)-ethanone; 3,4-dimethoxy benzoic acid, methyl ester; and (3) syringyl compounds: 1,2,3-trimethoxybenzene; 1,2,3-trimethoxy-5-methyl-benzene; 1-(3,4,5-trimethoxyphenyl)-ethanone; 3,4,5-trimethoxy-benzoic acid methyl ester. All standards were prepared

14

with eicosane as the internal standard and were run on the same day as the samples using identical instrumental parameters.

The products were quantified using an internal standard approach and by using either select ions or select ion ranges to measure peak areas. Relative response factors (RRF) were calculated for these standard compounds relative to eicosane. The RRF were then applied to these compounds produced during the TMAH reaction with the DOM samples to determine their concentrations (yields), which were normalized per mg organic carbon (OC). Knowing the DOC contents of the water, one could also report concentrations normalized to water volume. Sometimes, RRF were found to exhibit a linear dependence on concentration with a non-zero intercept. The RRF and concentrations for the data presented here were calculated using equation (1) that accounted for variations in the intercept, which were sometimes found to be substantial. The non-zero intercepts result from the method the software uses to calculate peak areas and do not affect the accuracy or precision of the quantitative results. Slight differences between RRF used for the Mount Rainier sample and the DOM samples result from differences in instrumental parameters and the select ion or mass range used to calculate the RRF.

$$\frac{A_i}{A_{IS}} = RRF\frac{Wt_i}{Wt_{IS}} + b \tag{1}$$

The terms in equation (1) are as follows: A_i and A_{IS} are the peak areas for the analyte and the internal standard, respectively, RRF is the relative response factor for the standard i, Wt_i and Wt_{IS} are the weights (ng) of the analyte and internal standard, respectively, and b is the correction factor (intercept). Two methods were used to calculate the RRF. For analytes with authentic standards, select ions (m/z) in the mass spectra were used to calculate peak areas, as this allowed for an increased signal to noise as compared to using the sum of some mass range. Select ions were chosen so that the m/z with the greatest relative intensity was used that also was not present in the spectra for co-eluting compounds.

TMAH products without authentic standards were also quantified. For the FAMEs observed in TMAH data without authentic standards, it is safe to assume that the intensity and fragmentation of these FAMEs would be similar to that of the neighboring FAMEs in the chromatograms for which standards existed. Therefore, for these FAMEs, the RRF calculated using the select ions were used for the quantifications.

For the aromatic compounds without authentic standards, the selection of a representative RRF was more complex. In cases where the identified DOM products were an isomer of one of the standards, RRF factors of the structural isomers were assumed to be the same as that of the standard compounds. Other aromatic compounds without authentic standards and that were not structural isomers of one of the standard compounds were also quantified. For these compounds, we used the RRF for standards that had similar structures and functional groups and that had similar retention times. However, for aromatic compounds without authentic standards, we could not use select ions for quantification. Here peak areas and RRF were calculated using the total ion current (TIC), which is the sum of all digital counts within a defined m/z range produced by the detected fragment ions. Using this GC-MS system, the TICs for co-eluting compounds can be deconvoluted by the Leco software to yield accurate and precise calculations for peak areas. RRF calculated with a deconvoluted select ion or TIC will be denoted with a D in front of the select ion or m/z range of the TIC in this report and the RRF used to quantify the TMAH products will also be presented. In the absence of pure reference standards, other studies have used a single RRF based on an average of a few lignin standards (Hatcher et al. 1995). We determined that this could cause errors as great as 80%, as typically RRF vary between 0.2 and 1.8.

^{13}C-Labeled TMAH Procedure

^{13}C TMAH was synthesized as described by Filley, Minard, and Hatcher (1999). Approximately 100 ± 2 mg of total dissolved solids (~1 mg C) were weighed into 2 separate glass ampoules (50 mg to each) along with 15 mg of solid ^{13}C TMAH and 100 μL of deionized, carbon filtered, and sterile filtered water (Elgastat UHQ MK II®, Elga, Bucks, England). The use of methanol was avoided because methanol would contribute unlabeled methoxyl groups to the products (Filley et al. 2000; Filley, Minard, and Hatcher 1999). The added water was then evaporated on the vacuum line. The ampoules were flame sealed and treated identically to the standard TMAH thermochemolysis procedure as described previously.

Identification of the ^{13}C labeled TMAH products was based on comparison of retention times with those of standards and unlabeled TMAH products from the same DOM sample. Product identifications were also verified by mass spectral analysis. Structural mass spectrometry was used to determine the amount of ^{13}C enrichment due to methylation. The equations used to determine the ^{13}C enrichment of the TMAH products were modified from Filley, Minard, and Hatcher (1999) so that the yield of the singly, doubly, or triply labeled products could be determined as opposed to percent demethylation.

The yield (ng) and percent composition of the product with one, two, or three ^{13}C methyl groups were calculated for selected TMAH products. All methoxy-substituted TMAH products were isotopically enriched with at least one ^{13}C TMAH methyl group and, therefore, the amounts of product that were unlabeled (not isotopically enriched) were not calculated. This means there was no methoxy-substituted TMAH product that was not initially a hydroxyl or, if there were natural methoxy groups, these were accompanied by at least one hydroxyl. $\%^{13}C_{SL}$ is the percent of singly labeled TMAH product, $\%^{13}C_{DL}$ is the percent of the compound that acquired two ^{13}C labels, and $\%^{13}C_{TL}$ is the percent of compound that acquired three ^{13}C labels. The abundance of each of these species was calculated using equations 2, 3, and 4 respectively. ML, ML_2, and ML_3 are the abundances of the molecular ions for the singly labeled, doubly labeled, and triply labeled TMAH products, respectively. Because the unlabeled carbons in a compound contain natural abundances of carbon isotopes with a 1.1% relative abundance of ^{13}C, this contribution

$$\%^{13}C_{SL} = \frac{ML}{(ML + ML_2 + ML_3)} \cdot 100 \tag{2}$$

$$\%^{13}C_{DL} = \frac{ML_2}{(ML + ML_2 + ML_3)} \cdot 100 \tag{3}$$

$$\%^{13}C_{TL} = \frac{ML_3}{(ML + ML_2 + ML_3)} \cdot 100 \tag{4}$$

$$ML_2 = I_{m+2} - \frac{n_L}{n}(ML+1)_{calc} \qquad (5)$$

$$ML_3 = I_{m+3} - \frac{n_L}{n}(ML_2+1)_{calc} \qquad (6)$$

$$(ML+1)_{calc} = \frac{m+1}{m} \cdot ML \qquad (7)$$

$$ML_{ald} = \frac{m}{m-H}(ML_{ald}-1) \qquad (8)$$

to a $m+1$ intensity can influence the measurement of Ml_i and a correction is required. ML_2 and ML_3 were calculated using equations 5 and 6, which were necessary to correct for natural ^{13}C abundance of ML and ML_2, respectively. I_{m+2} is the relative abundance of the m/z + 2 greater than the unlabeled TMAH products molecular ion and I_{m+3} likewise. Also, n_L is the number of natural carbons for the labeled compound (not including the labeled methoxy groups), whereas n is the number of carbons for the corresponding unlabeled TMAH product (including carbons from all methoxy functionalities). These terms are included to correct for the additional ~1.1% per methoxy of ^{13}C from the unlabeled TMAH that is not seen in the corresponding labeled compound (see Filley et al. (2000) for an in depth description). The abundance of the $(ML+1)_{calc}$ is calculated by determining the ratio of natural ^{13}C abundance $(m+1)$ relative to the molecular ion m from the unlabeled TMAH product and then by multiplying this ratio by ML (equation 7). $(ML_2+1)_{calc}$ is calculated in the same manner by substituting ML_2 for ML in equation 7.

For 3,4-dimethoxy benzaldehyde (G4) a modified equation (8) was used to calculate ML (here Ml_{ald}) as a result of the substantial m-H peak of ML_2 that interferes with the ML_{ald} abundance determination. The ratio of m over m-H was determined from the unlabeled TMAH product. $(ML_{ald}-1)$ is the m-H abundance of the singly labeled TMAH product. For all other compounds the m-H peak was found to be insignificant and no correction was required. Unfortunately, lack of sufficient sample precluded ^{13}C TMAH analysis of the Rio Tempisquito DOM and Rio Tempisquito RDOM. Therefore, only results from the White Clay Creek DOM and RDOM are presented.

STATISTICAL ANALYSIS

For the T-RFLP analyses, the similarity of the types of fragments between samples was calculated using Sorensen's coefficient. Hierarchical clustering was performed on binary coded datasets using unweighted pair-group method (UPGMA) (McCune and Grace 2002) and displayed as a dendrogram (a diagram showing the evolutionary relationship of species) in PC-ORD (McCune 1999). Sampling regimes for bioreactors included duplicate bioreactors for most treatments and duplicate samples from each bioreactor. Environmental samples from sediment and epilithon included duplicate nucleic acid extractions from the same sample and multiple samples from each site. At least two samples from each site were analyzed. The samples used in the analysis for each objective are listed in the next section of the report. Previous work has shown that the coefficient of variation (c.v.) of the relative abundance of t-RFLP peaks associated with PCR is 17% and the c.v. associated with nucleic acid extraction is 15% (unpublished data). t-RFLP traces from replicate bioreactors showed that the peak patterns in each trace were very similar (see Figure 3.20A, p. 68). However, statistical analysis of the data was represented as presence or absence of peaks and therefore unaffected by c.v. associated with relative percent of total sample area.

DOC data were placed into SAS data sets (SAS Institute, Cary NC) maintained on a microVAX 3100 at the Stroud Water Research Center. Significant differences were determined with an alpha error level of $P = 0.05$. Relative precision was defined as the coefficient of variation for replicate analyses, expressed as a percent.

Reproducibility of the TMAH thermochemolysis reaction was evaluated with a lignin standard, and averaged 6.2% (range 0.4 to 24% for all compounds). When authentic standards, and thus select ions, were used to calculate yield, the variance ranged from 0.4% to 7.7%.

CHAPTER 3
RESULTS AND DISCUSSION

OBJECTIVE 1A. CHARACTERIZATION OF NOM AND BOM POOLS FROM DIFFERENT SOURCE WATERS

DOC concentrations in the 5 source waters exhibited marked temporal variations, typically associated with storms (Appendix A). Under baseflow conditions, concentrations for each site generally increased through the series, Rio Tempisquito (0.4 to 0.8 mg C/L), White Clay Creek (0.8 to 1.8 mg C/L), Barker Lake (2.0 to 3.0 mg C/L), White River (2.5 to 3.5 mg C/L), and Hillsborough River (3.0 to 5.0 mg C/L) (Table 3.1). The relative precision of the DOC measurements, or coefficient of variation, calculated as the product of the mean divided by the standard deviation and expressed as a percent, varied among the individual laboratories, but were consistently less than 10%. The coefficients of variation for the DOC analyses, separated by site, were as follows: White Clay Creek, 2.2%; Rio Tempisquito, 7.2%; Hillsborough River, 2.2%; White River, 0.7%; and Barker Lake, 6.1%.

The BDOC concentrations also varied temporally (Appendix A), and while we were unable to positively identify the influences of land use or vegetation on these data, we did observe that the deciduous forest sites had the highest BDOC concentrations, the tropical evergreen source had the lowest concentrations, and the alpine and coniferous/cypress sites were intermediate. Humic substances and carbohydrates were constituents of NOM sources, and for the White River water, we determined that both were also part of the BDOC pool. In fact, on three separate dates, between 8 to 16% of the humic-DOC and 81 to 91% of the carbohydrate-DOC were metabolized in the duplicate bioreactors labeled K_2 and K_{13} (Tables 3.2-3.4). TMAH thermochemolysis analyses of bioreactor inflow and outflow showed a reduction in numerous peaks for both the White River (Figure 3.1) and the Hillsborough River (Figure 3.2).

Table 3.1
DOC and BDOC concentrations (mg C/L) in source waters

Source (period of record)	DOC* (range)	BDOC* (range)	BDOC as % DOC* (range)
White Clay Creek (January 2000 to May 2001)	2.93 ± 2.93 (289) (0.78-17.26)	0.75 ± 0.52 (45) (0.11-3.08)	37.56 ± 12.02 (45) (10.30-74.70)
Río Tempisquito (January 2000 to May 2001)	0.74 ± 0.29 (399) (0.44-2.75)	0.14 ± 0.09 (69) (0.03-0.56)	21.00 ± 9.38 (69) (5.60-55.90)
White River (March 1999 to November 1999)	3.73 ± 0.74 (16) (2.63-5.04)	1.11 ± 0.22 (11) (0.81-1.67)	29.87 ± 3.79 (11) (21.73-34.85)
Barker Lake (March 2002 to November 2002)	2.79 ± 0.75 (29) (1.71-5.17)	0.34 ± 0.47 (14) (0.00-1.78)	11.29 ± 11.67 (14) (0.00-41.78)
Hillsborough River (November 1999 to May 2000)	4.56 ± 1.73 (17) (2.80-10.00)	0.40 ± 0.22 (17) (0.00-0.80)	9.68 ± 5.78 (17) (0.00-18.92)

*Data expressed as $\bar{x}$ ± SD (n).

Table 3.2
White River water sampled at Indiana American 02 May 2000 for BOM characterization

Source	DOC (µg/L)	% Humic	DTCHO (µg/L)	BDOC (µg/L) [% DOC]
Raw	3786 ± 4	42.7	258	
Settled	2308 ± 14	39.5	78	
Filtered	2205 ± 18	45.0	72	
K_2 bioreactor outflow	2609 ± 5	51.4	24	1180 [31%]
K_{13} bioreactor outflow	2622 ± 6	51.8	23	1167 [31%]

Table 3.3
White River water sampled at Indiana American 02 June 2000 for BOM characterization

Source	DOC (µg/L)	% Humic	DTCHO (µg/L)	BDOC (µg/L) [% DOC]
Raw	3263 ± 15	45.3	111	
Settled	2168 ± 12	44.4	38	
Filtered	2080 ± 5	44.4	33	
K_2 bioreactor outflow	2653 ± 74	50.9	14	610 [19%]
K_{13} bioreactor outflow	2646 ± 5	51.5	21	617 [19%]

Table 3.4
White River water sampled at Indiana American 13 June 2000 for BOM characterization

Source	DOC (µg/L)	% Humic	DTCHO (µg/L)	BDOC (µg/L) [% DOC]
Raw	3076 ± 39	48.6	150	
Settled	2238 ± 24	44.5	69	
Filtered	2031 ± 13	41.9	41	
K_2 bioreactor outflow	2388 ± 29	48.5	19	688 [22%]
K_{13} bioreactor outflow	2646 ± 5	55.2	22	430 [14%]

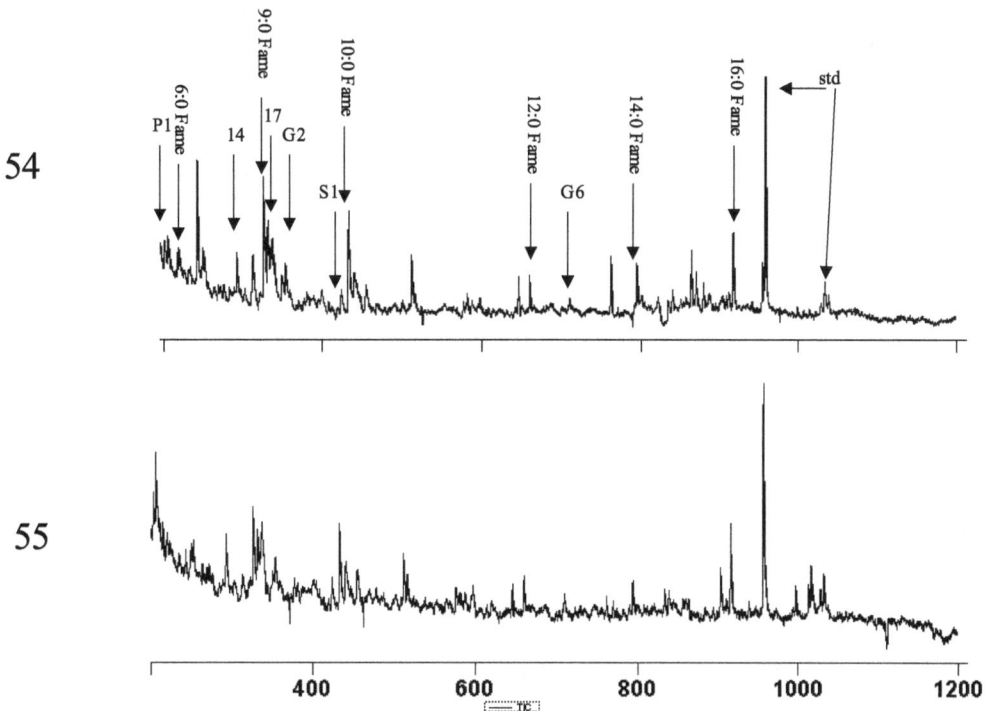

Figure 3.1 Chromatogram for TMAH thermochemolysis products from DOM from Indiana American water facility, including White River raw water (54) and bioreactor outflow (55)

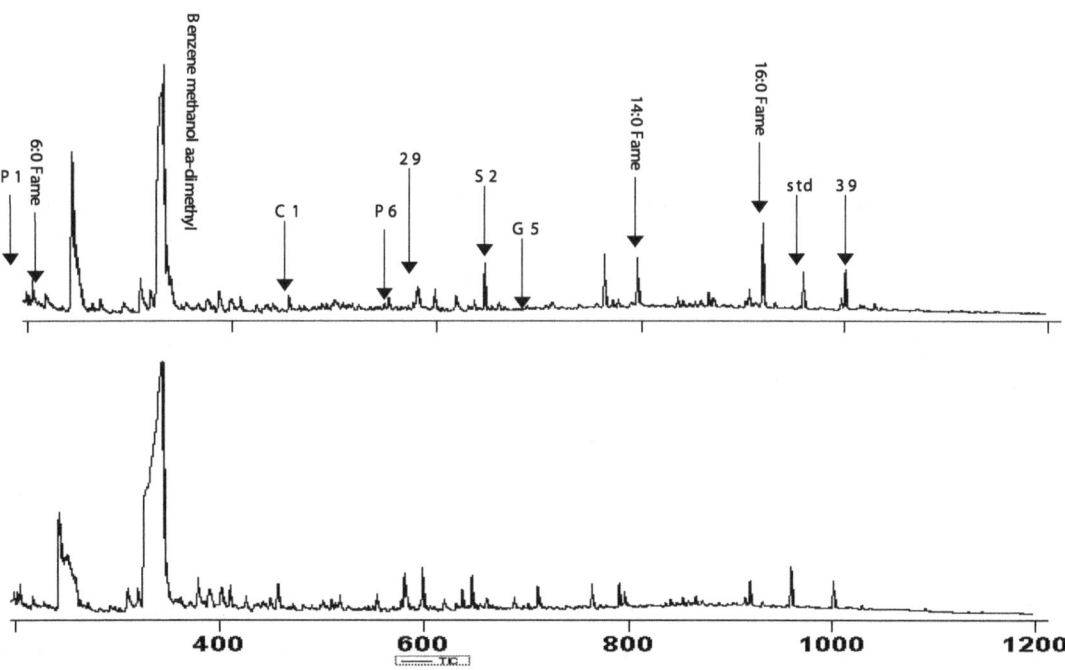

Figure 3.2 Chromatogram for TMAH thermochemolysis products from DOM in raw water from the Hillsborough River (upper) and bioreactor outflow (lower) in Tampa Bay water treatment plant

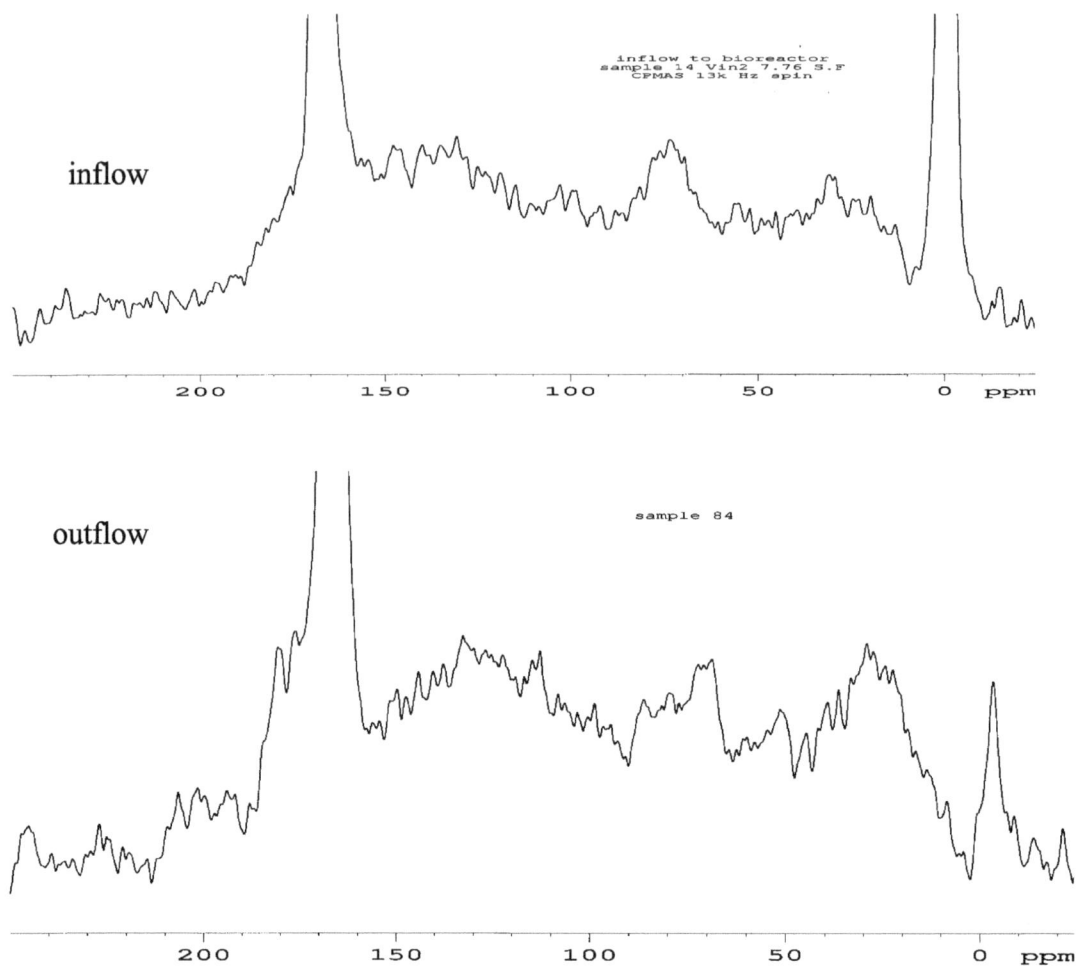

Figure 3.3 ^{13}C NMR spectra of DOM Rio Tempisquito water

Analyses of both NOM and BOM were derived from the application of solid-state ^{13}C NMR to the inflow and outflow to the bioreactor colonized by and operated on water from the Rio Tempisquito. The NMR analyses give an average composition for DOM by providing information on the nature of carbon functional groups in the sample. The general chemical shift regions associated with ^{13}C NMR spectra are paraffinic structures between 0-45 ppm, aliphatic C-N structures between 40-60 ppm, methoxyl structures between 50-58 ppm, carbohydrate related structures between 60-100 ppm, C-substituted aromatic structures between 100-140 ppm, carboxyl and amide related structures between 160-180, and aldehydes and ketone carbons between 180-220.

The NMR data for inflow and outflow water samples from the bioreactor associated with Rio Tempisquito are presented in Figure 3.3. Aside from the intense signals at about 0 and 170 ppm that are due to polysiloxanes from silicone tubing and bicarbonate from the salts, respectively, the NMR spectra indicate that the DOM is relatively enriched in aromatic signals (100-170 ppm) not unlike what we might expect from a material that has a significant contribution from lignin or lignin-derived carbon. The carbohydrate signals at 72 ppm seem to be reduced in relative intensity in the bioreactor effluent, indicating utilization in the bioreactor. It appears that the bioreactor effectively removes signals from the siloxanes, possibly by an adsorption process.

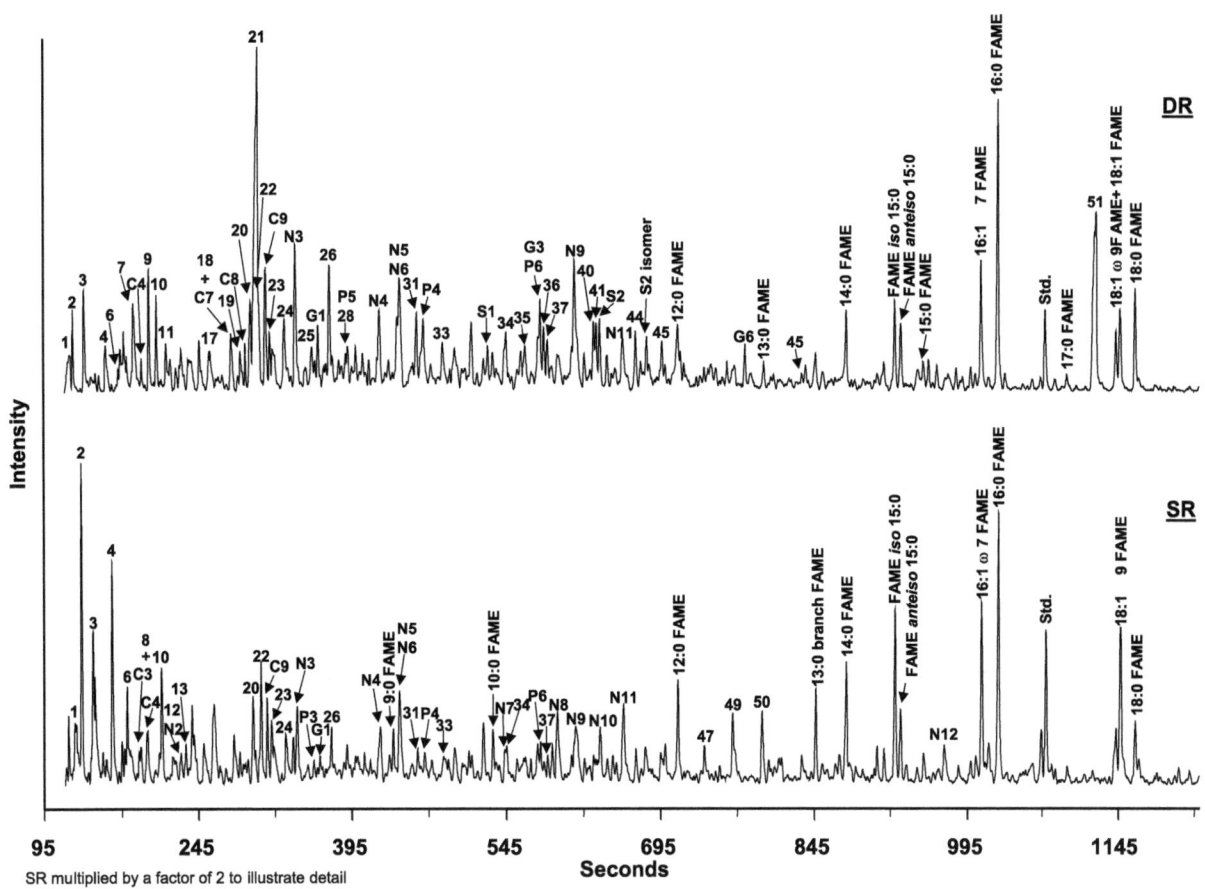

Figure 3.4 TMAH thermochemolysis chromatograms for DOM from Delaware and Schuylkill rivers

Qualitative Characterization of NOM and BOM with TMAH

Examples of the TMAH GC-MS chromatograms for two of the four DOM samples are shown in Figure 3.4. The peak identities for standards, mass quantitation, and response factors are listed in Table 3.5, while some of the commonly observed TMAH products of DOM and the peak identifications for the accompanying chromatograms are listed in Table 3.6. These are categorized under subheadings suggesting potential precursors—although there is sometimes overlap and uncertainty when assigning sources for TMAH products. In particular, many of the methoxy benzene TMAH products can result from proteins, lignin, carbohydrates, and tannins. However, there are a number of TMAH products that have distinct sources. For example, many of the carbohydrate- and lignin-derived products are unique to those sources. The following discussion is intended to describe some of the origins for the observed TMAH products in detail.

Products of Uncertain Origin

Many of the TMAH products of uncertain origin in Table 3.6 appear to be specific to particular, but presently unknown, bio-molecules. These compounds vary with the source of DOM

23

Table 3.5
List of standards used for TMAH analysis, their quantitation masses, and response factors

Peak	Compound	Quant. Mass*	Surrogate standard†	RRF‡	b‡	r²	%RSD RRF§
Mt. Rainier degraded lignin							
1	benzyl methyl ether	D50:200	P2	1.78			1.06
3	2-methoxy phenol	D50:200	G1	1.35			3.57
	benzoic acid, methyl ester	D50:200	G1	1.35			3.57
G1	1,2-dimethoxy benzene	95		0.590	-0.0795	0.999	
G2	3,4-dimethoxytoluene	77		0.305	-0.0666	0.999	
P4	4-methoxy-benzaldehyde	135		0.519	-0.272	0.999	
S1	1,2,3-trimethoxybenzene	110		0.399	-0.119	0.999	
G3	4-ethenyl-1,2-dimethoxy benzene	D50:200	G2	1.12			3.60
5	1,2,4-trimethoxy benzene	D50:200	S1	1.07			4.53
G20	3,4-dimethoxy benzyl ethe	D50:200	G5	0.637			8.95
G4	3,4-dimethoxy benzaldehyde	95		0.271	-0.156	0.999	
G21	1,2-dimethoxy-4-(1-propenyl) benzene	D50:200	G22	0.575			11.1
6	1,2-dimethoxy4-n-propylbenzene	D50:200	G22	0.575			11.1
G5	1-(3,4-dimethoxyphenyl)-ethanone	165		0.367	-0.222	0.998	
G22	1-(3,4-dimethoxyphenyl)-methyl ethanone	151		0.593	-0.529	0.998	
G6	3,4-dimethoxy benzoic acid, methyl ester	196		0.242	-0.168	0.999	
G23	2-methoxy-1-(3,4-dimethoxyphenyl) propane	D50:200	G22	0.575			11.1
7	dimethoxy-4-ethyl benzaldehyde	D50:200	G5	0.637			8.95
G24	3,4-dimethoxy benzeneacetic acid, methyl ester	D50:200	G6	1.00			7.76
8	2,5-dimethoxy-4-ethyl benzaldehyde	D50:200	G5	0.637			8.95
G10 or G11	cis or trans-1-(3,4-dimethoxyphenyl)-1-methoxy-1-propene	D50:200	G5	0.637			8.95
S5	1-(3,4,5-trimethoxyphenyl)-ethanone	195		0.238	-0.181	0.998	
G12	3,4-dimethoxy benzenepropanoic acid, methyl ester	D50:200	S6	0.631			7.62
9	dimethoxy benzenepropanoic acid, methyl ester	D50:200	S6	0.631			7.62
DOM samples							
22	benzoic acid, methyl ester	D35:200	G1	1.14	-0.0950	0.999	
8:0 FAME	octanoic acid, methyl ester	74		2.04	-0.196	0.997	
P3	1-ethenyl-4-methoxy-benzene	D35:200	P4	0.683	-0.164	0.998	
G1	1,2-dimethoxy-benzene	77		0.644	-0.0681	0.999	
26	1,4-dimethoxy-benzene	D35:200	G1	1.14	-0.0950	0.999	
31	2,5-dimethoxytoluene	D35:200	G2	0.890	-0.128	0.999	
	dimethoxytoluene	D35:200	G2	0.890	-0.128	0.999	
	3,4-dimethoxytoluene	77		0.299	-0.0492	0.999	

(continued)

Table 3.5 (Continued)

Peak	Compound	Quant. Mass*	Surrogate standard†	RRF‡	b‡	r^2	%RSD RRF§
P4	4-methoxy-benzaldehyde	135		0.442	-0.162	0.999	
33	benzenepropanoic acid, methyl ester	D35:200	P4	0.683	-0.164	0.998	
S1	1,2,3-trimethoxybenzene	110		0.359	-0.0918	0.998	
10:0 FAME	decanoic acid, methyl ester	74		1.61	-0.517	0.992	
34	3-methoxy benzoic acid, methyl ester	D35:200	P6	1.04	-0.209	0.997	
P5	1-(4-methoxy phenyl) ethanone	135		0.584	-0.260	0.996	
G3	4-ethenyl-1,2-dimethoxy-benzene	D35:200	G4	0.593	-0.100	0.999	
P6	4-methoxy-benzoic acid, methyl ester	135		0.962	-0.326	0.996	
36	1,2,4-trimethoxybenzene	D35:200	S1	0.937	-0.233	0.999	
S2	1,2,3-trimethoxy-5-methyl-benzene	53		0.465	-0.110	0.998	
40	3-methoxy-4-methylbenzoic acid, methyl ester	D35:200	P6	1.04	-0.209	0.997	
G4	3,4-dimethoxy-benzaldehyde	77		0.231	-0.0738	0.999	
44	2-methoxy-benzenepropanoic acid, methyl ester	D35:200	G6	0.938	-0.238	0.998	
12:0 FAME	dodecanoic acid, methyl ester	74		1.33	-0.264	0.996	
G5	1-(3,4-dimethoxyphenyl) ethanone	77		0.159	-0.0400	0.999	
G6	3,4-dimethoxy benzoic acid, methyl ester	165		0.346	-0.116	0.997	
S5	1-(3,4,5-trimethoxyphenyl)-ethanone	195		0.206	-0.0622	0.998	
14:0 FAME	tetradecanoic acid, methyl ester	74		1.11	-0.159	0.997	
S6	3,4,5-trimethoxy-benzoic acid, methyl ester	59		0.141	-0.0422	0.997	
FAME iso-C_{15}	2-methyl tetradecanoic acid, methyl ester	D74	14:0 FAME	1.11	-0.159	0.997	
FAME anteiso-C_{15}	3-methyl tetradecanoic acid, methyl ester	D74	14:0 FAME	1.11	-0.159	0.997	
16:1 ω7 FAME	9-hexadecenoic acid, methyl ester	55		0.386	-0.0170	0.999	
16:0 FAME	hexadecanoic acid, methyl ester	74		1.09	-0.234	0.998	
18:1 ω9 FAME	9-octadecenoic acid, methyl ester	84		0.092	-0.0050	0.991	
18:0 FAME	octadecanoic acid, methyl ester	74		0.830	-0.0990	0.998	
20:0 FAME	eicosanoic acid, methyl ester	74		0.617	-0.0530	0.999	
22:0 FAME	docosanoic acid, methyl ester	74		0.532	-0.0370	0.999	
24:0 FAME	tetracosanoic acid, methyl ester	74		0.409	-0.0340	0.999	
26:0 FAME**	hexacosanoic acid, methyl ester	74		0.260	-0.0208	—	

* where select ions were inappropriate the sum of m/z between #:# were used. Where adjacent peaks shared the quantitation m/z a deconvolution algorithm was used to determine individual peak areas based. The deconvoluted m/z was signified by D#.

† where authentic standards were not available the closest eluting standard with similar structure was used as an estimate for RRF.

‡ The RRF, b, and r^2 were calculated using the equation $(A_a/A_{is})=RRF\ (C_a/C_{is})+b$ where A_a and A_{iS} were the peak areas for the analyte and internal standard respectively, C is ng, and b is the intercept.

§ where RRF were calculated using average RRF from varying concentrations of analyte the % relative standard deviation (%RSD RRF) was used to assess the variance. This method was used when b approached zero.

** since authentic standards were not obtained RRF and b were extrapolated from the linear trend observed for FAMES 18:0 through 24:0.

Table 3.6
List of identified products from TMAH thermochemolysis
of DOM samples and Mount Rainier lignin

Peak	Compound
TMAH products of uncertain origin	
1	acetamide
2	3-methoxy pentane
3	2-hydroxy-3-methyl butyric acid, methyl ester
4	2-methoxy-3-methyl butyric acid, methyl ester
5	methyl pentanoic acid, methyl ester
6	4-methoxy butanoic acid, methyl ester
7	1,1'-oxybis[2-methoxy]-ethane
8	benzaldehyde
9	α-methyl styrene
10	benzyl methyl ether
11	4-oxo pentanoic acid, methyl ester
12	1-methoxy-2-methyl benzene
13	decahydro naphthalene
14	1-methoxy-3-methyl benzene
15	methoxy butanoic acid, methyl ester
16	2-ethyl hexanol or related isomer
17	butanedioic acid, dimethyl ester
18	acetophenone
19	methyl butanedioic acid, dimethyl ester
20	N,N-dimethyl benzamine
21	α,α-dimethyl benzenemethanol
22	benzoic acid, methyl ester
23	methyl decahydro naphthalene
24	unknown compound (m/z 42, 127, 58)
25	pentanedioic acid, dimethyl ester
26	1,4-dimethoxy benzene
27	methyl decahydro naphthalene
28	methyl pentanedioic acid, dimethyl ester
29	benzeneacetic acid, methyl ester
30	unknown compound (m/z 42, 127, 58, 142)
31	1,4-dimethoxy-2-methyl benzene
32	hexanedioic acid, dimethyl ester

(continued)

26

Table 3.6 (Continued)

Peak	Compound
33	benzenepropanoic acid, methyl ester
34	3-methoxy benzoic acid, methyl ester
35	4-ethyl-1,2-dimethoxy benzene
36	1,2,4-trimethoxy benzene
37	3-phenyl-2-propenoic acid, methyl ester
38	methoxy benzeneacetic acid, methyl ester
39	1,3,5-trimethoxy benzene
40	3-methoxy-4-methyl benzoic acid, methyl ester
41	2,6-di-tert-butyl-4-methyl phenol (BHT)
42	octanedioic acid, dimethyl ester
43	1,2,3-trimethoxy-5-(2-propenyl) benzene or isomer
44	2-methoxy benzenepropanoic acid, methyl ester
45	methoxy benzenepropanoic acid, methyl ester
46	nonanedioic acid, dimethyl ester
47	methyl diphenyl methyl ether
48	3-(4-methoxyphenyl) 2-propenoic acid, methyl ester
49	hexadecene
50	benzophenone
51	1,1'-(1-methylidene) bis-4-methoxy benzene

Carbohydrate derived TMAH products

C1	2-methyl-2-cyclopenten-1-one
C2	3,4-dimethyl-2-cyclopenten-1-one
C3	dihydro 3-methyl-2(3H) furanone
C4	dihydro 4-methyl-2(3H) furanone
C5	tetrahydro 2H-pyran-2-one
C6	methoxy-dimethyl cyclohexenone
C7	2,3,4-trimethyl-2-cyclopenten-1-one
C8	1-cyclohexyl ethanone
C9	2-ethyl-3-methoxy-2-cyclopenten-1-one
C10	1,5-anhydro-2,3,4,6-tetra-o-methyl-d-mannitol

Ligin derived TMAH products

P1	methoxy benzene
P3	1-ethenyl-4-methoxy benzene

(continued)

Table 3.6 (Continued)

Peak	Compound
G1	1,2-dimethoxy benzene
P5	1-(4-methoxy phenyl) ethanone
P4	4-methoxy benzaldehyde
S1	1,2,3-trimethoxy benzene
G3	4-ethenyl 1,2-dimethoxy benzaldehyde
P6	4-methoxy benzoic acid, methyl ester
S2	1,2,3-trimethoxy-5-methyl benzene
G4	3,4-dimethoxy benzaldehyde
G5	1-(3,4-dimethoxy phenyl) ethanone
G6	3,4-dimethoxy benzoic acid, methyl ester
S6	3,4,5-trimethoxy benzoic acid, methyl ester
G24	3,4-dimethoxy benzeneacetic acid, methyl ester

Fatty acid, methyl esters (FAMEs)

8:0 FAME	octanoic acid, methyl ester
9:0 FAME	nonanoic acid, methyl ester
10:0 FAME	decanoic acid, methyl ester
12:0 FAME	dodecanoic acid, methyl ester
13:0 branch FAME	methyl tridecanoic acid, methyl ester
14:0 FAME	tetradecanoic acid, methyl ester
FAME *iso* 15:0	2-methyl tetradecanoic acid, methyl ester
FAME *anteiso* 15:0	3-methyl tetradecanoic acid, methyl ester
15:0 FAME	pentadecanoic acid, methyl ester
16:1 ω 7 FAME	9-hexadecenoic acid, methyl ester
16:0 FAME	hexadecanoic acid, methyl ester
17:1 FAME	heptadecenoic acid, methyl ester
17:0 FAME	heptadecanoic acid, methyl ester
18:1 FAME	octadecenoic acid, methyl ester
18:1 ω 9 FAME	9-octadecenoic acid, methyl ester
18:0 FAME	octadecanoic acid, methyl ester

Alkanes

Alk 15	C_{15} alkane
Alk 16	C_{16} alkane

(continued)

Table 3.6 (Continued)

Peak	Compound
Alk 17	C_{17} alkane
pristane	C_{19} branched alkane (pristane)
Alk 18	C_{18} alkane
phytane	C_{20} branched alkane (phytane)
Alk 19	C_{19} alkane

N containing compounds

N1	1-methyl-1H-pyrrole-2-carboxaldehyde
N2	1-methyl-2,5-pyrrolidinedione
N3	unknown heterocyclic N compound (m/z 56,141,156)
N4	3-ethyl-1,3-dimethyl-2,5-pyrrolidinedione
N5	3,5-dimethyl-1-propylpyrrole
N6	4,6-dimethyl-3,5-dioxo-2,3,4,5-tetrahydro triazine
N7	1,3-dimethyl-3,4,5,6-tetrahydro-2H pyrimidinedione
N8	1,3,5-trimethyl-1,3,5-triazine-2,4,6-trione
N9	1,3-dimethyl-2,4(1H,3H) pyrimidinedione
N10	5,6-dimethyl-1H benzothiazole
N11	1,3,5-trimethyl-2,4(1H,3H) pyrimidinedione
N12	caffeine

that presumably has unique inputs of organic matter and is, therefore, likely specific to one or few biochemical sources. Some such compounds are acetamide (peak 1); α-methyl styrene (9); decahydro naphthalene (13); α,α-dimethyl benzenemethanol (21); methyl decahydro naphthalene (23 and 27); some diacid methyl esters (32, 42, and 46); benzophenone (51); and 1,1'-(1-methylidene) bis-4-methoxy benzene (52). Some benzenepropanoic and benzenepropenoic acid methyl esters (33, 37, 44, 45, and 49) may be protein-derived (Manino and Harvey 2000; Zang et al. 2000). However, many products seem to be general products of the TMAH reaction with organic compounds. Many of these products have been observed in studies involving various organic matter types including pure reference compounds (McKinney et al. 1996; Chefetz et al. 2000a, b; Mannino and Harvey 2000; del Rio et al. 1998; Fabbri and Helleur 1999). We believe that the following compounds are TMAH products that are less specific in origin with regard to particular bio-molecules and appear ubiquitously as TMAH products from DOM: 3-methoxy pentane (2); the butanoic, butyric, and pentanoic acid methyl esters (3-6, 11, 15, 17, 25, and 28); benzyl methyl ether (10); N,N-dimethyl benzamine (20); benzoic acid methyl ester (22); peak 24; the dimethoxy benzenes (G1 and 26); peak N3; peak 30; the trimethoxy benzenes (S1, 36, and 39); 4,6-dimethyl-3,5-dioxo-2,3,4,5-tetrahydro triazine (N6); 1,3,5-trimethyl-1,3,5-triazine-2,4,6-trione (N8); and caffeine (N12). However, there is a benefit in including these compounds in comparisons of DOM since the relative peak areas for these compounds do vary between DOM samples and can, thus, be useful for "fingerprinting."

Carbohydrates

Generally, the early retention times in the chromatograms are enriched in carbohydrate-derived compounds. The four DOM samples have similar distributions of carbohydrate-derived compounds. The White Clay Creek DOM exhibits some higher molecular weight carbohydrate-derived compounds (e.g. C10 and others not illustrated), which are minor constituents of the chromatogram and cannot be precisely identified by comparison to our mass spectral libraries. However, these are likely to be more useful for characterizing the carbohydrate nature of the DOM than the general carbohydrate products in the earlier retention times (Fabbri and Helleur 1999). Several methoxy benzene compounds are also observed as TMAH products from carbohydrates. These include 1,2- and 1,4-dimethoxy benzene; 1,2,3- and 1,2,4-trimethoxy benzene; and trimethoxy toluenes [unpublished results; Fabbri and Helleur 1999).

Lignin

Although not predominant peaks, lignin-derived products resulting from the three different lignin subunits (p-hydroxyphenyl (P), guaiacyl (G), and syringyl (S)) are observed in these DOM samples. It appears that not all of the lignin products listed in Table 3.6 are necessarily specific only to lignin. For example, 1,2-dimethoxy benzene (G1); 1,2,3-trimethoxybenzene (S1); and 1,2,3-trimethoxy-5-methyl-benzene (S2) can derive from other biopolymers such as carbohydrates and tannins [unpublished results], whereas methoxy benzene (P1) and 4-methoxy benzoic acid, methyl ester (P6) can derive from proteins (Ertel, Hedges, and Perdue 1984). The distribution of lignin-derived compounds appears to depend on the source of the DOM and, thus, is related to the unique inputs from the indigenous vegetation and the unique, in-stream processes for each of the sampled waters. Although each DOM sample has its own unique distribution of lignin-derived components, all DOM samples share a common feature in that there are no lignin-derived TMAH products that have a side chain greater than a 2-carbon unit. Lignin products with side chains greater than a 2-carbon unit are commonly encountered as TMAH products of pure lignin or wood samples (see Mount Rainier below, Filley, Minard, and Hatcher 1999). It is likely that lignin structures are substantially degraded in natural waters, such that they produce few or no TMAH products with extended side chains. It is known that microbes attack the lignin polymers and selectively oxidize the α-carbon in the side chains (Crawford 1981). Being that the lignin-derived material in natural waters is potentially in an advanced degree of degradation, a large proportion of the α-carbon may be oxidized and cleaved from the rest of the lignin biopolymer resulting in increased relative amounts of these P, G, and S compounds with no extended side chains.

Aliphatic Products

Fatty acid methyl esters (FAMEs) are the most predominant TMAH products observed in the chromatograms of the DOM samples. FAMEs have both microbial and plant origins and range in size from the octanoic acid methyl ester (8:0 FAME) through tetratricontanoic acid methyl ester (not illustrated), where hexadecanoic acid methyl ester (16:0 FAME) and octadecanoic acid methyl ester (18:0 FAME) are often the most prominent peaks. These FAMEs are accompanied by several mono-unsaturated and branched-chain isomers. Although each of the DOM samples has a complete series of FAMEs, the distributions or yields of the individual FAMEs are unique for each sample (Figure 3.4). The relative proportions of the lower-molecular weight FAMEs (8:0 through 10:0 FAME) compared to the higher molecular weight FAMEs (12:0 through 18:0 FAME) and the proportion of unsaturated vs. saturated FAMEs varies between samples. The differences in the FAME distributions likely reflect the unique inputs from and effects of the indigenous microbial communities.

A series of alkanes, varying in size from pentadecane (Alk 15) through nonadecane (Alk 19) accompanied by the branch chain isomers pristane and phytane, is observed as TMAH products from the Rio Tempisquito DOM sample (Figure 3.4, Table 3.6). These alkanes are likely algal-derived since they are a comparatively lower molecular weight series, peaking at C_{17}, than expected for terrestrially derived plant waxes. Also, both pristane and phytane are present and likely come from phytol, a diagenesis product of the diterpenoid alcohol side chain of chlorophyll (Meyers et al. 1984). Another interesting set of aliphatic compounds encountered in these data is the decahydronaphthalenes and methyl decahydronaphthalenes (peaks 13, 23, and 27). These compounds can potentially result from the aliphatic-alicyclic components or moieties described by Leenheer (1994).

Nitrogen Containing Compounds

Heterocyclic-N containing and other N containing compounds are found in the early to mid retention times, but these are not major components of DOM as represented in the chromatograms. The origins of many of these heterocyclic-N containing compounds are presently unclear. Some of the N-containing compounds may be protein-derived whereas others have been identified as non-source specific products of the TMAH procedure such as the compound corresponding to peak N3; 4,6-dimethyl-3,5-dioxo-2,3,4,5-tetrahydro triazine; 1,3,5-trimethyl-1,3,5-triazine-2,4,6-trione; and caffeine (see discussion below). In general, the amino acid methyl esters that result from proteins are difficult to identify as a result of the complexity of these chromatograms and their low yields, although valine- and leucine-derived methyl esters (Zang et al. 2000) can be identified as minor components in a few of these DOM samples.

Recovery of Added Standards

As described above, an experiment was designed to determine the optimal point during the TMAH procedure for adding internal standards. We observed a substantial amount of degradation or possibly polymerization (thus removing compounds from detection) for many of the lignin standards spiked into the DOM sample prior to the TMAH reaction. This was evident by comparisons of the peak areas for chosen reference products that originate from the DOM (different products than those in the standard mixture, e.g. the 16:0 FAME) relative to the peak areas of the individual spiked standards (Figure 3.5). The lignin standards were also degraded to a greater degree than eicosane in the pre-reaction standard addition results. This result was expected since eicosane is less reactive than aromatic compounds with activating functional groups. In the case where standards were spiked into the ampoules prior to the TMAH reaction, substantially decreased recoveries (measured by peak areas) were observed for the standards when compared to the case where standards were added after the TMAH reaction and prior to extraction. However, the degree to which these compounds were degraded was variable, depending upon their structure. For example, 1,2-dimethoxy benzene (G1); 3,4-dimethoxy toluene (G2); 1,2,3-trimethoxy benzene (S1); 3,4,5-trimethoxy toluene (S2); and eicosane were comparatively stable relative to

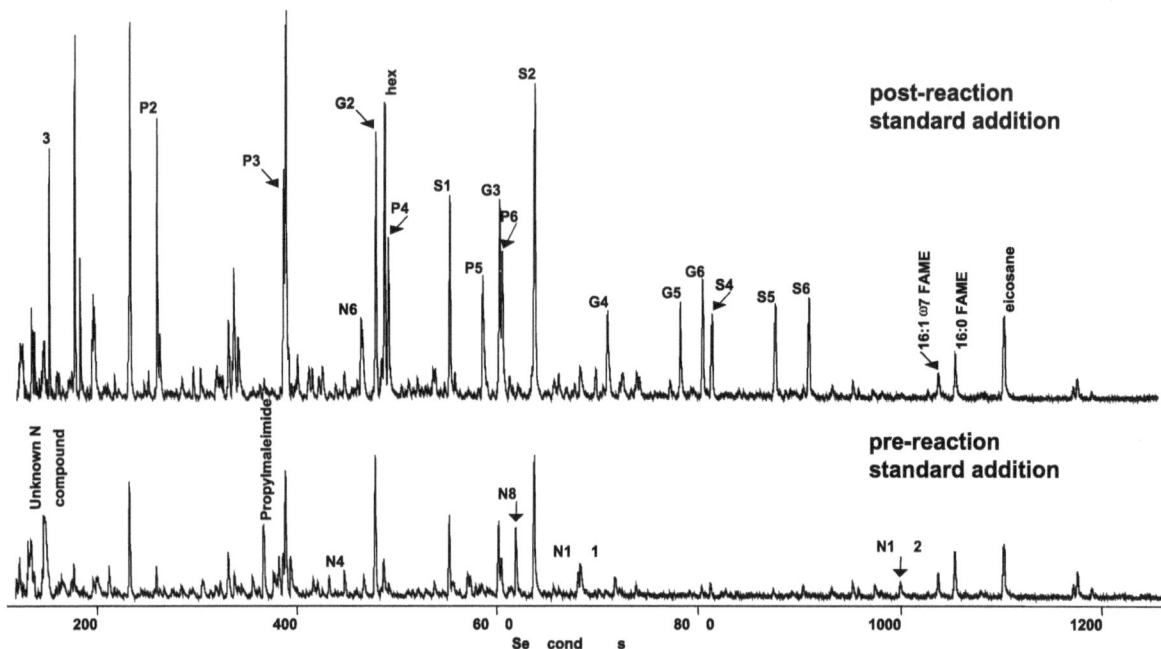

Figure 3.5 TMAH results for a DOM sample to which a set of standards was added prior to (below) or after (above) the TMAH reaction

the other lignin standards that were degraded to a greater degree. Therefore, we conclude that adding internal standards prior to the TMAH reaction can lead to substantial errors and increased experimental variance in quantitative results. For this reason, we have chosen to add the internal standard after the TMAH reaction, but prior to product extraction. Adding standards in this sequence, we obtained nearly complete recovery relative to eicosane (within five percent) for each of the lignin standards. This indicates that sorption of the standards to the residual DOM or mineral matrix is negligible and that the choice between which of the standards are used as the internal standard(s) is insignificant, provided they do not overlap with sample-derived products.

Increased yields of some heterocyclic N-containing compounds were observed in the results for the pre-reaction standard addition samples as compared to the post-reaction standard addition samples. These compounds are an unidentified N-containing compound; propyl-maleimide (some related isomer of); 3-ethyl, 1,3-dimethyl-2,5- pyrrolidinedione (N4); 1,3,5-tri-methyl-1,3,5-triazine-2,4,6-trione (N8); 1,3,5-trimethyl-2,4-pyrimidinedione (N11); and caffeine (N12) (Figure 3.5). This indicates that methoxy-aromatic compounds are degraded by the TMAH reaction and in this process heterocyclic-N containing products are produced as degradation products. Using these as indicators of naturally derived heterocyclic-N containing molecules or proteins may lead to misinterpretations.

TMAH Thermochemolysis Results for the Mt. Rainier Lignin

The TMAH GC-MS chromatogram for the Mt. Rainier sample of degraded lignin, produced with the sum of m/z between 50 and 150, is shown in Figure 3.6 with the accompanying peak identifications and quantified yields for the TMAH products presented in Table 3.7. Most commercially available lignin standards involve harsh chemical treatment for isolation and offer

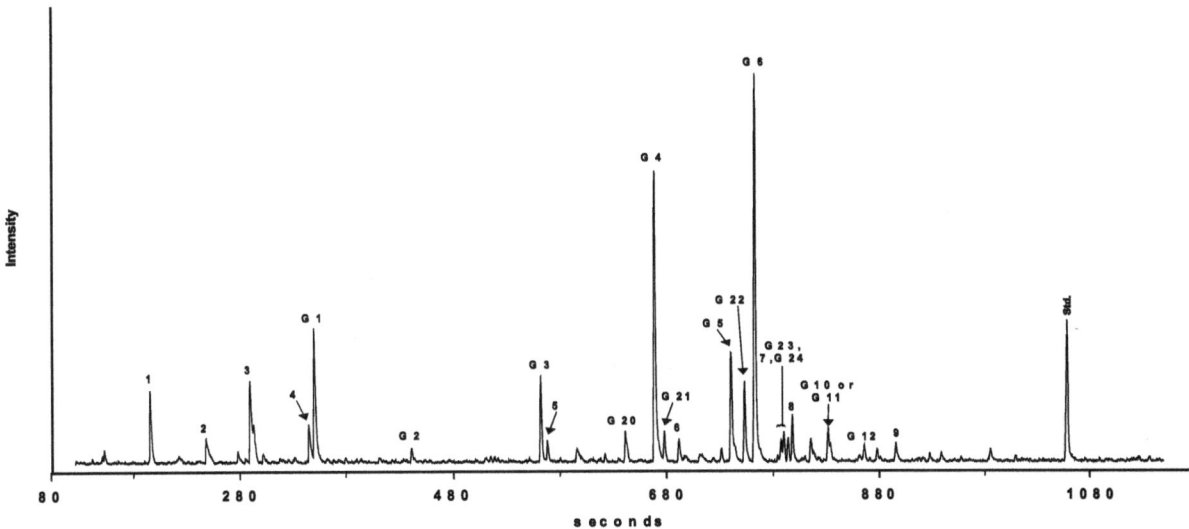

Figure 3.6 TMAH chromatogram of Mt. Rainier degraded lignin sample produced with the sum of m/z between 50 and 150

no advantages over the natural standard we have used that relies upon microbial processes to remove the cellulosic materials from wood. There is little evidence supporting repolymerization even though it was suggested some 100 years ago. Most likely, oxidized lignin is simply washed out of the wood.

The chromatogram compares well with previously published data (Hatcher et al. 1995). The predominance of guaiacyl (G) compounds is expected from gymnosperm lignin and these compounds outweigh other compounds in terms of yield by 1-2 orders of magnitude. The other observed TMAH products, those that are not dimethoxy benzene derivatives, are general products that result from the TMAH thermochemolysis reaction with organic compounds as they have been observed in studies involving various organic matter types including pure reference compounds.

One purpose of this experiment was to evaluate the reproducibility of the TMAH thermochemolysis reaction. The reproducibility for the TMAH procedure was on average 6.2% and varied between 0.4% and 24% for all compounds. When authentic standards, and thus select ions, were used to calculate the yield, the variance ranged between 0.4% and 7.7%. Therefore, we conclude that pure reference standards are required to obtain precise quantitative results.

Ratios, such as the $(Ad/Al)_G$ ratio commonly used to represent the degree of degradation for lignin, can be more confidently compared between samples, for example, that are run on different instruments or at different times by using absolute concentrations since response factors (peak areas) fluctuate. Often $(Ad/Al)_G$ can range from < 1 to greater than 10 for guaiacyl and syringyl phenol derivatives (Hatcher et al. 1995). The $(Ad/Al)_G$ for this degraded wood sample was 1.03 ± 0.05 (mean, SD) when calculated using absolute concentrations and was 1.17 ± 0.06 when using peak areas integrated using the TIC (sum of m/z 35-500) (Table 3.7). The larger value calculated using the TIC partially results from a few, small peaks co-eluting with G6 and by the difference in the relative response factors (RRFs) for G4 and G6 (0.756 and 1.00 respectively).

Though this TMAH procedure only marginally yields products from cellulose, the absence of any carbohydrate-derived compounds shows that little cellulose is present in the sample. Therefore, if we assume that this sample is purely lignin we can calculate the total yield of

Table 3.7
Concentrations of TMAH thermochemolysis products from Mount Rainier lignin

Peak	Compound	Conc. ng (mg OC^{-1})	SD	%RSD
1	benzyl methyl ether	510	24.1	4.73
2	butanedioic acid dimethyl ester			
3	2-methoxy phenol	1330	293	22.0
4	pentanedioic acid dimethyl ester			
	benzoic acid, methyl ester	222	11.0	4.95
G1	1,2-dimethoxy benzene	1770	56.1	3.17
G2	3,4-dimethoxytoluene	326	3.87	1.19
P4*	benzaldehyde, 4-methoxy-	545	2.05	0.376
S1*	1,2,3-trimethoxybenzene	310	8.50	2.74
G3	4-ethenyl-1,2-dimethoxy benzene	1050	37.7	3.59
5	1,2,4-trimethoxy benzene	406	9.92	2.44
G20	3,4-dimethoxy benzyl ether	876	83.2	9.50
G4	3,4-dimethoxy benzaldehyde	5820	79.6	1.37
G21	1,2-dimethoxy-4-(1-propenyl) benzene	935	94.9	10.1
6	1,2-dimethoxy4-n-propylbenzene	736	16.7	2.27
G5	1-(3,4-dimethoxyphenyl)-ethanone	3787	290	7.66
G22	1-(3,4-dimethoxyphenyl)-methyl ethanone	3360	19.2	0.571
G6	3,4-dimethoxy benzoic acid, methyl ester	6000	223	3.72
G23	2-methoxy-1-(3,4-dimethoxyphenyl) propane	766	184	24.0
7	dimethoxy-4-ethyl benzaldehyde	971	44.9	4.62
G24	3,4-dimethoxy benzeneacetic acid, methyl ester	713	66.9	9.38
9	2,5-dimethoxy-4-ethyl benzaldehyde	1650	117	7.09
G10 or G11	cis or trans-1-(3,4-dimethoxyphenyl)-1-methoxy-1-propene	1350	59.7	4.42
S5	ethanone, 1-(3,4,5-trimethoxyphenyl)	376	3.80	1.01
G12	3,4-dimethoxy benzenepropanoic acid, methyl ester	388	10.4	2.68
11	dimethoxy benzenepropanoic acid, methyl ester	584	89.3	15.3
std.	eicosane			
	Total	34781	291	0.837
	% of lignin represented	3.3	0.2	6.06
	(Ad/Al)$_G$ (calculated using absolute concentration)	1.03	0.05	5.05
	(Ad/Al)$_G$ (calculated using integrated areas)	1.17	0.06	5.13

* Peaks not labeled in chromatograph.

TMAH products and estimate the fraction of lignin represented by the TMAH procedure on a per weight basis. The yield of quantifiable compounds from this lignin is 3.3% ± 0.2 (mean%, SD) (Table 3.7). Although this fraction may seem small, the efficiency of the TMAH reaction is on the order of other degradative techniques that commonly represent <10% of the organic matter (Ertel, Hedges, and Perdue 1984; Hatcher et al. 1995).

These discussed advantages and confidence in comparisons can be realized for degradative GC-MS methods only where the internal standard technique can be properly implemented. In addition, standard compounds are essential for precise quantitative comparisons. Unfortunately, not all TMAH products of organic matter are easily accessible and we recognize the need for synthesizing standard compounds to develop a more complete set. Nevertheless, we consider the reproducibility of this TMAH procedure satisfactory for quantitative analysis of environmental samples.

Quantitative Yields of TMAH Products From DOM

The quantitative TMAH thermochemolysis GC-MS analyses of the DOM samples are presented in Table 3.8. Due to limited sample size, we were unable to replicate the TMAH results for the DOM samples. However, we wanted to demonstrate that the quantitative TMAH thermochemolysis procedure is useful for investigating DOM molecular composition. Therefore, we chose the Mount Rainier reference standard to establish the quantitative reproducibility of the TMAH procedure. The product yields are normalized to organic carbon (OC) for the individual TMAH products, using the elemental analysis results, as opposed to volume (ng L^{-1}) so that compositional differences in the nature of DOM could be investigated. The percent organic carbon by weight of the isolated dissolved materials from Rio Tempisquito (RT), White Clay Creek (WCC), Delaware River (DR), and Schuylkill River (SR) were 1.04%, 1.85%, 2.67%, and 1.02%, respectively. These percents are low because salts contribute substantially to the weight of the total dissolved solids (TDS).

Table 3.8 is organized as follows: total quantified aromatic contribution (ng (mg OC)$^{-1}$) is the sum of the yields for all aromatic products including the potentially lignin-derived compounds since there is some overlap in these products with those from other sources. The total quantified lignin contribution is the sum of the yields for only the lignin-derived, aromatic TMAH products (therefore, the yield of non-lignin aromatic products is the difference between total aromatic yield and the total lignin yield). The total quantified FAME yield is the sum of the yields for all FAMEs. The percent quantified aromatic compounds relative to total yield of products is calculated by subtracting the weight of lignin-derived, aromatic compounds from the total quantified aromatic contribution (only non-lignin aromatic products are represented) and dividing by the total yield of all TMAH products. The percent of lignin-derived products and FAMEs relative to total yield is calculated by dividing the total product yield for lignin-derived products by the total yield of all TMAH products (Table 3.8).

The ratio of the total quantified lignin products: non-lignin, aromatic products: FAME was consistently ~1:1:1.8 with the exception of the Delaware River sample which was 1:1:1, where the concentration of the lignin-derived products and the aromatic products of non-lignin origin slightly exceeded that of FAMEs. The DOM samples have similar total yields of products, with the exception of the Rio Tempisquito sample, in which the product yields are roughly twice that of the other samples. This indicates that a greater proportion of the DOM is represented in the Rio Tempisquito sample using TMAH thermochemolysis GC-MS as compared to the other DOM

Table 3.8
Concentrations of TMAH thermochemolysis products for DOM samples

Peak	Compound	Concentrations (ng (mg OC⁻¹))			
		RT	WCC	DR	SR
22	benzoic acid, methyl ester	633	235	175	298
8:0 FAME	octanoic acid, methyl ester	200	61.1	39.4	116
P3	1-ethenyl-4-methoxy-benzene	187	212	166	190
G1	1,2-dimethoxy-benzene	428	228	157	88.9
26	1,4-dimethoxy-benzene	456	420	300	250
31	2,5-dimethoxytoluene	253	75.4	77.2	189
	dimethoxytoluene	301	94.1	113	133
	3,4-dimethoxytoluene	166	182	209	104
P4	4-methoxy-benzaldehyde	418	231	244	182
33	benzenepropanoic acid, methyl ester	194	140	130	203
S1	1,2,3-trimethoxybenzene	493	205	152	54.8
10:0 FAME	decanoic acid, methyl ester	394	148	104	285
34	3-methoxy benzoic acid, methyl ester	268	124	149	195
P5	1-(4-methoxy phenyl) ethanone	248	144	152	142
G3	4-ethenyl-1,2-dimethoxy-benzene	310	234	207	137
P6	4-methoxy-benzoic acid, methyl ester	424	195	278	223
36	1,2,4-trimethoxybenzene	874	341	257	175
S2	1,2,3-trimethoxy-5-methyl-benzene	141	182	47.7	55.1
40	3-methoxy-4-methylbenzoic acid, methyl ester	362	205	261	238
G4	3,4-dimethoxy-benzaldehyde	476	187	307	286
44	2-methoxy-benzenepropanoic acid, methyl ester	221	137	248	252
12:0 FAME	dodecanoic acid, methyl ester	362	181	121	270
G5	1-(3,4-dimethoxyphenyl) ethanone	310	47.6	88.7	14.7
G6	3,4-dimethoxy benzoic acid, methyl ester	951	138	213	60.2
S5	1-(3,4,5-trimethoxyphenyl)-ethanone	135	25.1	17.1	28.5
14:0 FAME	tetradecanoic acid, methyl ester	615	152	150	308
S6	3,4,5-trimethoxy-benzoic acid, methyl ester	363	64.8	44.6	45.9
FAME *iso* 15:0	2-methyl tetradecanoic acid, methyl ester	169	109	109	252
FAME *anteiso* 15:0	3-methyl tetradecanoic acid, methyl ester	370	259	89.5	152
16:1 ω 7 FAME	9-hexadecenoic acid, methyl ester	1150	1230	320	655
16:0 FAME	hexadecanoic acid, methyl ester	1937	454	512	688
18:1 ω 9 FAME	9-octadecenoic acid, methyl ester	323	555	111	596
18:0 FAME	octadecanoic acid, methyl ester	1280	93.7	184	159
20:0 FAME	eicosanoic acid, methyl ester	77.6	17.3	39.8	46.8
22:0 FAME	docosanoic acid, methyl ester	85.9	15.8	65.4	68.5
24:0 FAME	tetracosanoic acid, methyl ester	114	19.4	74.2	61.3
26:0 FAME	hexacosanoic acid, methyl ester	55.6	10.9	71.5	35.6
	Total aromatic contribution (ng (mg OC⁻¹))	8612	4047	3993	3545
	Total lignin contribution (ng (mg OC⁻¹))	4884	2094	2074	1508
	Total FAME contribution (ng (mg OC⁻¹))	7078	3295	1919	3658
	% of aromatic compounds of total	23.8	26.6	32.5	28.3
	% of lignin compounds of total	31.1	28.5	35.1	20.9
	% of FAME of total	45.1	44.9	32.5	50.8
	Estimated lignin contribution (% by weight)	14.8	6.4	6.3	4.6
	(Ad/Al)$_G$	2.00	0.74	0.69	0.21
	16:0 FAME / 26:0 FAME	34.8	41.7	7.2	19.3
	(*iso* 15:0 + *anteiso* 15:0) / 16:0	0.28	0.81	0.39	0.59

samples. The yield of non-lignin aromatic products from the Schuylkill River DOM sample is similar to the other DOM samples. However, the lignin product yields are markedly lower in the Schuylkill River sample compared to the other DOM samples, indicating lower lignin inputs or more complete degradation of lignin within this river.

Bulk characteristics of DOM might be estimated or compared using TMAH data. For example, by dividing the total lignin yield (ng (mg OC)$^{-1}$) by the TMAH yield calculated for degraded lignin (3.3% for Mount Rainier), one can calculate the percent lignin composition of the DOM. This is based on the assumption that all lignin contributions between samples exhibit a constant yield of TMAH products and that this yield is the same as that of a pure degraded lignin sample. This is a tenuous assumption because it is likely that the lignin in DOM has undergone in-stream transformations that would affect product yields, and thus bias such an estimate. Encouragingly, the product yield for a base (NaOH) extract of the Mount Rainier sample, which presumably resembles a more degraded form of lignin, was similar to that of the non-extracted material (3.71%, data not shown). The results of these calculations show that the lignin contribution varies considerably between these samples and suggest that lignin may contribute 4.6-15% of the weight of these DOM samples (Table 3.8). Interestingly, the Rio Tempisquito is estimated to have 15% lignin whereas the other three samples are estimated to have 6% or less of lignin. This is consistent with the results from NMR studies mentioned previously above. Therefore, TMAH thermochemolysis GC-MS is useful for distinguishing differences in lignin contributions to DOM.

TMAH thermochemolysis is unique compared to other degradative techniques in that one can simultaneously derive biogeochemical information from fatty acids and lignin. In addition, Mannino and Harvey (2000) recently demonstrated that TMAH thermochemolysis yields similar results for the fatty acid analysis of DOM as the conventional solvent extraction techniques, though TMAH is much faster and easier. Therefore, we are confident in interpreting TMAH-derived data for fatty acids in the same manner, as data from the conventional lipid analyses (Meyers and Ishiwatari 1993; Napolitano 1999).

A few examples of relationships that have been developed for the analysis of DOM using fatty acids are presented in Table 3.8. The hexadecanoic acid (16:0):hexacosanoic acid (26:0) ratio has been suggested to indicate the relative amounts of microbial to terrestrial contributions (from higher plants) of organic matter (Meyers et al. 1984). This is based on the observations that microbial communities do not produce significant amounts of FAMEs >18:0 and, thus, these higher molecular weight FAMEs are most likely derived from cuticular material, bark, and suberin of terrestrial vegetation. It is assumed that plant contributions are not substantially affected by the diagenesis of organic matter since these larger, saturated fatty acids are somewhat resistant to degradation relative to the lower molecular weight fatty acids. The low values for the 16:0 to 26:0 ratios here suggest negligible amounts of terrestrial input to the DOM. However, the Delaware River and Schuylkill River appear to have more terrestrial input than Rio Tempisquito or White Clay Creek. The (*iso* 15:0 + *anteiso* 15:0) / 16:0 ratio has been used to estimate the relative contributions of prokaryotic to eukaryotic sources of FAMEs (Napolitano 1999). These results suggest that bacterial contributions of fatty acids are more substantial than those of eucaryotes for these samples (Table 3.8).

The following discussions concentrate on the yields for the individual TMAH products, which are given in Table 3.8. Each sample presented here has unique distributions and yields of FAMEs. For instance, the unsaturated FAMEs are more predominant than the saturated counterparts in White Clay Creek, whereas the inverse is true for the other DOM samples. For all samples except White Clay Creek the 16:0 FAME is the predominant FAME. For the White Clay Creek sample the 16:1 ω7 FAME is the predominant FAME. Although the Delaware River sample does not have the highest FAME yield as compared to the FAME distribution of the other samples, more of its FAME yield is distributed among the higher molecular weight FAMEs. Thus, the Delaware River sample may contain similar higher plant contributions, but less microbial contributions of fatty acids to the DOM as compared to the other DOM samples.

There are some commonalities between the FAME distributions for all four DOM samples. The FAME distributions for these DOM samples indicate predominant bacterial contributions of fatty acids. This is supported by the substantial presence of the *iso* 15:0 and *anteiso* 15:0 FAMEs, lack of poly-unsaturated fatty acids (PUFAs), enhanced concentrations of the 16:1 ω7 and 18:1 ω9 fatty acids, and the relatively low amounts of FAMEs >18:0 (Table 3.8). The possibility that PLFAs were selectively and quickly degraded could also explain their absence since they are extremely labile towards microbial degradation (Meyers et al. 1984; Napolitano 1999). Interestingly, although the Rio Tempisquito and the Delaware River samples result from very disparate environments, they have a similar distribution of FAMEs (though the Rio Tempisquito DOM has higher FAME yield).

Also illustrated in Table 3.8 are the yields for the aromatic TMAH products. In general, the yields for the aromatic products are comparable between the White Clay Creek, Delaware River, and Schyulkill River DOM samples, but are substantially greater for the Rio Tempisquito DOM sample, particularly for the lignin-derived aromatic products. This indicates that the Rio Tempisquito DOM sample is more amenable to the TMAH thermochemolysis reaction, which is consistent with the NMR data that show a higher aromatic carbon content than observed for White Clay Creek samples. The p-hydroxyphenyl TMAH products have similar distributions in terms of relative product yields for each of the DOM samples, though p-hydroxyphenyl yields are greater for the Delaware River and Schuylkill River DOM samples. The relative distributions of guaiacyl compounds are similar for the DOM samples herein, with the exception of the Rio Tempiquito DOM sample. The Rio Tempisquito DOM sample yielded substantially greater amounts of 3,4-dimethoxy benzoic acid methyl ester (G6) compared to other guaiacyl compounds than did the other DOM samples. Yields for syringyl TMAH products are low for each of the DOM samples. In general, although subtle differences are observed, the relative distribution of lignin-derived products is similar between these DOM samples—though absolute yields may vary.

Substantial yields of carbohydrate-derived aromatic products (1,4-dimethoxy benzene; 1,2,3-trimethoxy benzene; and 1,2,4-trimethoxy benzene) are observed for all DOM samples, which is consistent with the large number and yields of other carbohydrate-derived products observed in the TMAH thermochemolysis GC-MS chromatograms (Figs. 3.7) and the NMR data. The increased yields of the benzene propanoic acid methyl esters, potentially protein-derived TMAH products (Mannino and Harvey 2000), relative to other aromatic products and the substantial yields of heterocyclic-N products in the chromatogram for the Schuylkill River DOM sample as compared to the other samples may indicate that the Schuylkill River DOM sample was relatively enriched in proteins. Interestingly, the yields of the aromatic products for the Delaware River and Schuylkill River compounds are remarkably similar, aside from benzoic acid methyl ester and 3,4-dimethoxy benzoic acid methyl ester.

TMAH thermochemolysis GC-MS is demonstrated herein to be effective for investigating the lignin, carbohydrate, and lipid signature of aquatic DOM. Furthermore, the quantitative ability of this procedure adds a new dimension to the comparison of DOM samples within and between laboratories. Such comparisons are difficult and subjective using conventional GC-MS techniques that are semi-quantitative at best. If the sources and yields of TMAH products of unknown origin can be determined, then they could be combined with products of known molecular origin to provide bulk compositional characteristics of DOM. The comparison of data gathered by various analytical techniques such as TMAH GC-MS, ^{13}C and ^{1}H NMR, and more sophisticated mass spectroscopy techniques could provide a more accurate and comprehensive picture of DOM. This advance in the structural understanding of DOM can expand efforts in molecular modeling of DOM and modeling of environmental processes in which DOM is involved. However, to proceed with such investigations involving TMAH thermochemolysis, supporting knowledge must be established through the study of pure compounds and isolated biomolecules. This will allow markers for particular biomolecules and the yields of TMAH products from different biomolecules to be established through the quantitative application of TMAH GC-MS. One caveat is that pure compounds may not entirely reflect the character of biomolecules encountered in natural waters, since natural processes can modify these biomolecules.

TMAH Thermochemolysis of DOM From Bioreactor Samples Collected From Rio Tempisquito and White Clay Creek

Dissolved organic matter (DOM) present in freshwater ecosystems originates from the secretion, degradation, and transformation of allochthonous and autochthonous biomolecules. Considering the multitude of biological sources and various processes that can transform molecules in DOM, it is not surprising that DOM becomes an extremely complex mixture of molecules spanning a broad range of molecular weights. DOM is an important, sizeable, and dynamic compartment of the global carbon cycle, but the molecular composition of DOM is presently poorly characterized (Hedges et al. 2000; Hedges, Keil, and Benner 1997; McCarthy, Hedges, and Benner 1998). In stream ecosystems, which have organic matter budgets dominated by allochthonous inputs, most of the organic matter processing is the result of heterotrophic respiration (Findlay et al. 1997). The metabolism of DOM by heterotrophic bacteria is considered to be the base of stream ecosystems from which energy and nutrients are transferred to higher trophic levels. This process has been referred to as the microbial loop (Pomeroy 1974). However, validating models of the microbial loop and measuring the fluxes of energy between the different compartments of the microbial loop have proven elusive and the conventional perception of these processes has recently been questioned (Thomas 1997).

Understanding both the composition and quantities of DOM that are available for heterotrophic metabolism are necessary to further develop our understanding of stream ecology and sources of drinking water. Recent evidence from studies using plug-flow biofilm reactors (bioreactors) have revealed many of the bulk properties of the biodegradable fraction of DOM (BDOM). These studies suggest that a large part (60% to 90%) of the DOM pool in stream water is refractory to heterotrophic metabolism (Kaplan and Newbold 1995; Volk, Volk, and Kaplan 1997). Up to 75% of this BDOM in stream water can be classified using XAD fractionation as humic substances (HS) (Volk, Volk, and Kaplan 1997), which are thought to constitute the higher molecular weight and more hydrophobic portions of DOM (Aiken et al. 1979). Other studies have been conducted to assess microbial productivity when supported by certain biomolecules and humic isolates (Ellis et al. 1999; Volk, Volk, and Kaplan 1997), but the molecular composition of the BDOM that is available to the indigenous microbial communities of stream ecosystems is largely unexplored. The lack of knowledge concerning BDOM composition is in part a result of the limited availability of analytical methods that can be used to characterize complex mixtures such as DOM.

Table 3.9
DOC results for the Rio Tempisquito and White Clay Creek DOM and
refractory DOM (RDOM) samples

Sample	DOC (mg L^{-1})*	BDOC (mg L^{-2})
RT DOM	0.650	0.055
RT RDOM	0.595	
WCC DOM	3.75	1.02
WCC RDOM	2.73	

* variance < 5%

We believe the TMAH thermochemolysis GC-MS procedure will be valuable in studying the structural composition of BDOM when combined with the action of microorganisms associated with plug-flow bioreactors of the nature described in this report. A drawback to the TMAH procedure is that natural methoxy groups, such as those found in lignin, cannot be distinguished from those introduced during the TMAH reaction. This can be overcome by using the new ^{13}C-labeled TMAH thermochemolysis GC-MS procedure. ^{13}C TMAH thermochemolysis maintains the same degradative and derivatization characteristics as that of unlabeled TMAH (Filley, Minard, and Hatcher 1999). However, ^{13}C TMAH thermochemolysis relies on ^{13}C labeled methyl groups in TMAH as the methylating agent so that naturally occurring methoxy groups can be distinguished from those produced during the TMAH thermochemolysis procedure. The position of the labeled methoxy group (or the position of the natural phenolic or hydroxyl functionality) can often be determined by analysis of the mass spectral fragmentation patterns. Therefore, ^{13}C-TMAH thermochemolysis both stabilizes chemically and thermally labile functionalities, those potentially lost using other degradative techniques, and allows one to identify the structure prior to derivatization.

In this study, we combined TMAH thermochemolysis GC-MS, and ^{13}C-labeled TMAH thermochemolysis GC-MS to investigate the molecular composition of DOM and BDOM from two stream ecosystems: a temperate deciduous stream ecosystem and a tropical "evergreen" stream ecosystem. These contrasting stream ecosystems were chosen to see if different ecosystems produce unique DOM and BDOM molecular compositions.

DOC analyses were performed at the time of sampling and were used to measure the fraction of DOM utilizable by heterotrophic bacteria (Table 3.9). The DOC concentrations in the Rio Tempisquito at the time of sampling were low (0.6 to 1.0 mg L^{-1}) and within the range reported by Newbold et al. (1995) for this stream. The DOC concentrations in White Clay Creek were slightly higher than normal for this stream (Newbold et al. 1997) as a result of a recent storm. The BDOC for the Rio Tempisquito and the White Clay Creek were 8.5% and 27.2%, respectively, of the initial DOC (Table 3.9). This was initially interpreted to indicate that the Rio Tempisquito bioreactors were less effective at utilizing the indigenous DOM than at the White Clay Creek site, or that there were real differences in the quality of the DOM between these streams.

Qualitative Comparison of TMAH Products From the Rio Tempisquito DOM and RDOM

The qualitative differences between the Rio Tempisquito DOM and refractory DOM (RDOM) samples are evident in the TMAH thermochemolysis GC-MS chromatograms (Figure 3.7). p-xylene (1), benzyl methyl ether (8), and the 16:0 FAME are the most prominent

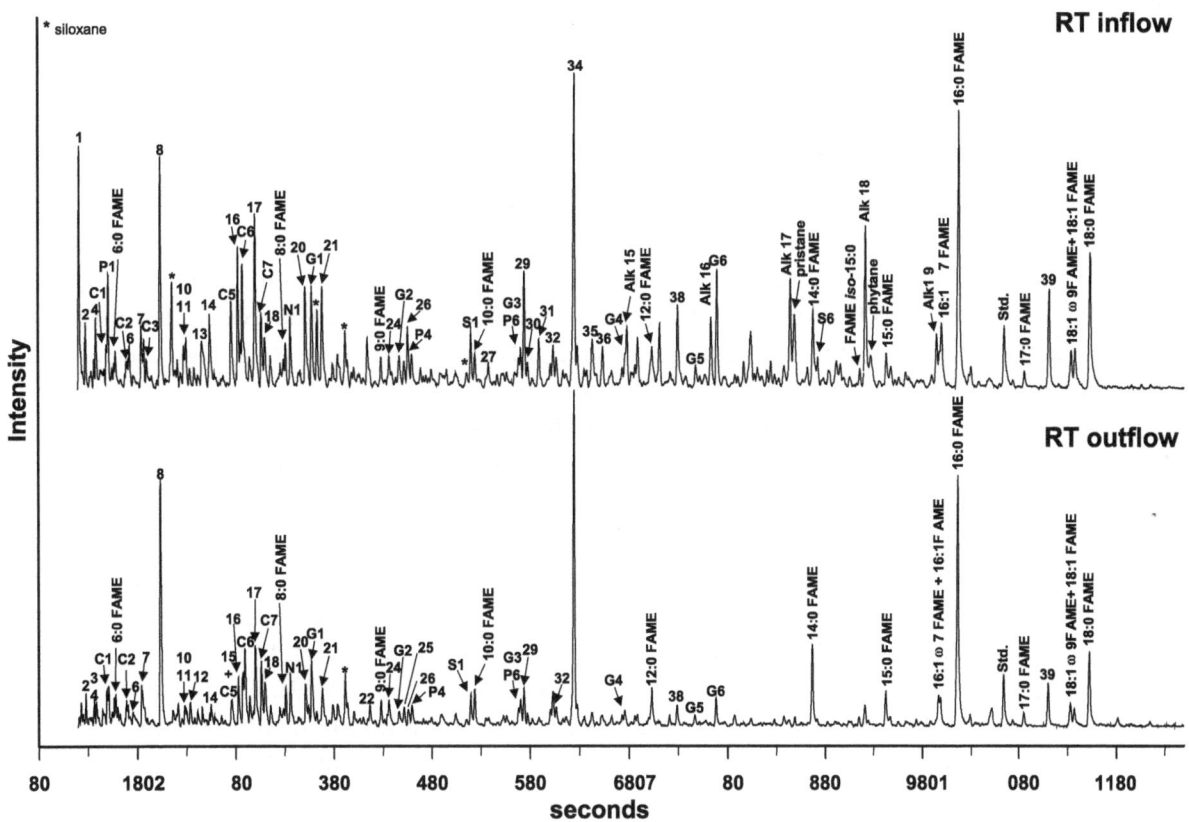

* Corresponds to a series of methyl siloxanes seen in the inflow DOM.

Figure 3.7 TMAH thermochemolysis GC-MS chromatograms for Rio Tempisquito (RT) bioreactors inflow and outflow DOM. Peak identifications are given in Table 3.5.

TMAH products from both the DOM and RDOM with the exception of p-xylene, whose precursor was completely degraded during passage through the bioreactors. BHT (34), 2,6-di-tert-butyl-4-methyl phenol, arises from the extraction solvent and is not a TMAH product of the DOM sample. Other peaks in the early to mid-retention times, that correspond mostly to carbohydrate and lignin-derived products as well as several products from unknown or non-specific sources, were reduced in the RDOM sample relative to the DOM sample, indicating that the precursors for these compounds contribute to the BDOM.

Methyl siloxanes were detected in the DOM, but not in the RDOM (Figure 3.7). The presence of siloxanes was verified by the presence of a –6 ppm peak in the CPMAS ^{13}C NMR spectra. These compounds may have been introduced during the sample collection procedure where silicon tubing was used. Notably, the removal of these compounds by the bioreactors was also supported with NMR data, and while the removal mechanisms are unknown, adsorption is probable.

Rio Tempisquito DOM and RDOM samples show a series of alkanes ranging from tetradecane to nonadecane, which shows no odd over even predominance. The relatively low molecular weight distribution of this series, centered around the C_{17} alkane, suggests algal/microbial sources as opposed to higher plant sources (Kawamura, Ishiwatari, and Ogura 1987; Meyers and Ishiwatari 1993). Interestingly, this alkane series was observed in DOM, but was completely removed during

41

degradation and is absent in the RDOM and thus contributes to the BDOM. This contradicts the notion that alkanes are relatively resistant to microbial degradation during these time scales (Meyers and Ishiwatari 1993).

Lower molecular weight alkanes have been suggested to degrade within the water column of lakes during sedimentation as well as within sediments, but this process takes considerably longer (Kawamura, Ishiwatari, and Ogura 1987; Meyers and Ishiwatari 1993; Meyers et al. 1984). Pristane and phytane were also observed and are characteristic of algal origins. These alkanes, thought to result from transformations of phytol, the isoprenoid side chain of chlorophyll a (Kawamura, Ishiwatari, and Ogura 1987; Meyers and Ishiwatari 1993), were degraded effectively by the bioreactors. This is surprising because these highly branched alkanes are also considered to be highly resistant to microbial degradation (Robson and Rowland 1988).

Throughout the chromatograms for both the DOM and RDOM samples, a series of FAMEs was observed. In general the distribution of FAMEs was similar between the DOM and RDOM samples and is indicative of bacteria since the distribution was dominated by saturated FAMEs with some monoenoic FAMEs (predominantly 16:1 ω7 and 18:1 ω9). Substantial amounts of *iso* and *anteiso* 15:0 FAMEs were also present, which suggests that there were contributions from gram-positive bacteria (Moll and Summers 1999). There was little contribution from eukaryotes as indicated by the lack of polyunsaturated fatty acids, though their absence could be expected since these fatty acids are quickly metabolized (Napolitano 1999). Contributions of fatty acids from higher plants were minimal as noted by the lack of higher molecular weight fatty acids (>C_{24} units). These results indicate that microbial contributions, primarily bacterial, to the DOM pool can be significant. A benzene dimer (1,1'-(1-methylidene) bis-4-methoxy benzene, peak 39) was observed in both the DOM and RDOM samples. This product derives from bisphenol A (4,4'-(1-methylethylidene)bis-Phenol (CAS 80-05-7)), which is used in the manufacturing of polycarbonate (see discussion of ^{13}C-labeled TMAH products below). Bisphenol A was likely introduced into these NOM samples during the sample collection procedure that included polycarbonate containers for storage and shipping of the stream water samples.

Qualitative Comparison of TMAH Products From the White Clay Creek DOM and RDOM

Substantial differences between the White Clay Creek DOM and RDOM samples were observed with TMAH thermochemolysis GC-MS, indicating that there were considerable contributions from BDOM to the product distribution (Figure 3.8). The dominant peaks were 3-methoxy pentane (2); benzyl methyl ether (8); 1,4-dimethoxy benzene (21); 1,3,5-trimethoxy benzene (32); and the 16:1 ω7 and 16:0 FAMEs. In the RDOM sample, these peaks were diminished with the exception of 3-methoxy pentane, benzyl methyl ether, and 2-methoxy-3-methyl butyric acid methyl ester (5), which were enhanced. All of these peaks, aside from the FAMEs, cannot be attributed to definite precursors at present.

As observed for the Rio Tempisquito DOM and RDOM, there was a reduction in peak area for most peaks in the early and mid-retention times corresponding to carbohydrate and lignin-derived products as well as other non-specific TMAH products. The peak corresponding to 2-ethyl-3-methoxy-2-cyclopenten-1-one (C7) was diminished to a lesser degree than the other peaks in these chromatograms. This indicates that the carbohydrate precursor for this product was more resistant to degradation within the bioreactors than the other forms of organic matter represented using TMAH thermochemolysis. Unlike the Rio Tempisquito samples, there were notable differences in the FAME distribution between the DOM and RDOM samples. For the White Clay

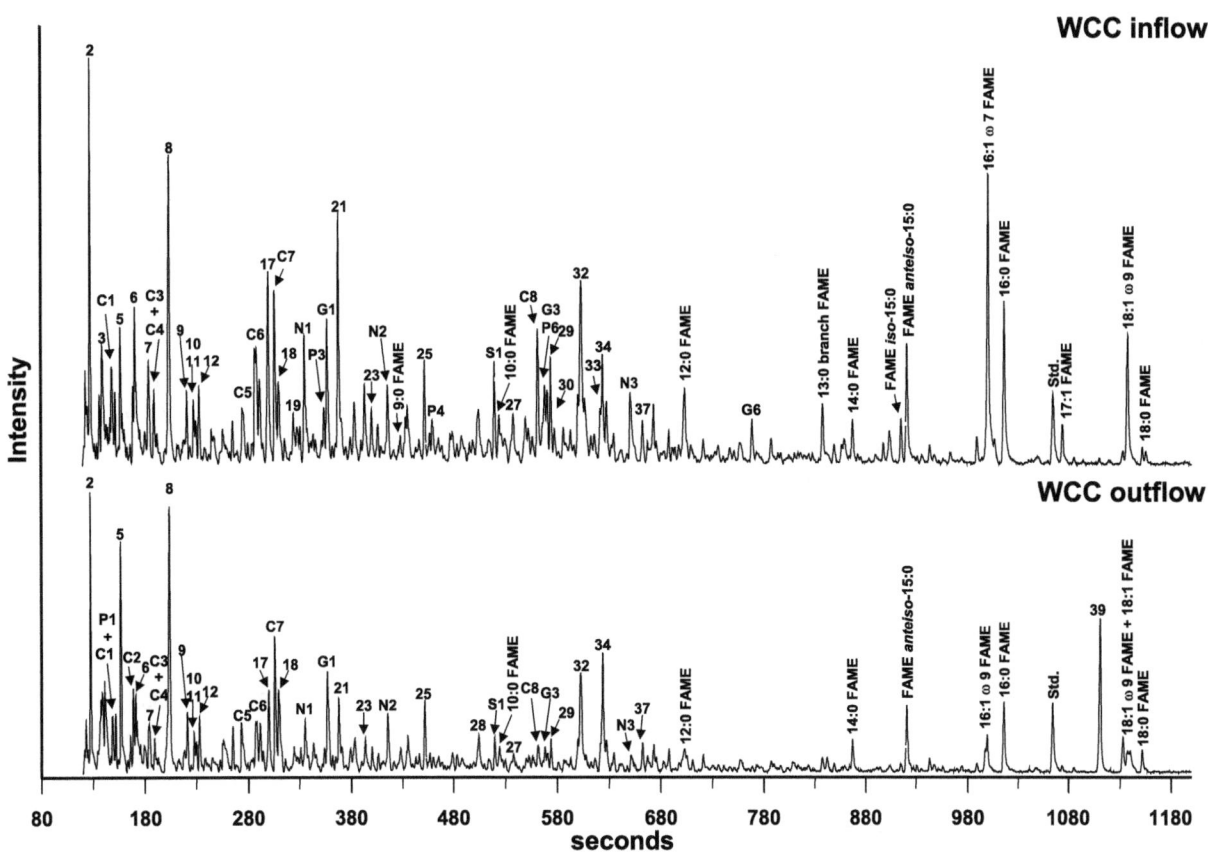

Figure 3.8 TMAH thermochemolysis GC-MS chromatograms for White Clay Creek (WCC) bioreactors inflow and outflow DOM. Peak identifications are given in Table 3.6.

Creek samples, there was a substantial reduction of peak area for all FAMEs with the most extreme reductions observed for the mono-unsaturated FAMEs. The FAME distribution suggests predominant bacterial contributions as observed for the Rio Tempisquito samples. 1,1'-(1-methylidene) bis-4-methoxy benzene (39) was also observed as a product from both the WCC DOM and RDOM samples.

Quantitative Comparison of TMAH Products From the Rio Tempisquito DOM and RDOM

Many TMAH products were quantified and they included lignin-derived aromatic and non-lignin derived, aromatic products as well as FAMEs for the Rio Tempisquito and White Clay Creek DOM and RDOM samples (Tables 3.10 and 3.11). Tables 3.10 and 3.11 are organized identically to Table 3.8, except that concentrations are expressed as ng L^{-1} rather than normalized to mass of organic C. The total quantified aromatic contribution (ng L^{-1}) is the sum of the yields for all aromatic products, including the potentially lignin-derived compounds. The total quantified lignin contribution is the sum of the yields for only the lignin-derived, aromatic TMAH products (therefore, the yield of non-lignin aromatic products is the difference between total aromatic yield and the total lignin yield). The total quantified FAME yield is the sum of the yields for all FAMEs. The percent quantified aromatic compounds relative to the total yield of products is calculated by

43

Table 3.10
Yields of TMAH products for Rio Tempisquito DOM and BDOM samples

Peak	Compound	Concentrations (ng L^{-1})			
		Inflow	Outflow	D ng	% Diff.
22	benzoic acid, methyl ester	449	136	313	70
8:0 FAME	octanoic acid, methyl ester	142	72	69.6	49
P3	1-ethenyl-4-methoxy-benzene	133	86	46.7	35
G1	1,2-dimethoxy-benzene	304	132	172	56
26	1,4-dimethoxy-benzene	324	93	231	71
31	2,5-dimethoxytoluene	180	63	117	65
	dimethoxytoluene	214	70	144	67
	3,4-dimethoxytoluene	118	56	62.0	53
P4	4-methoxy-benzaldehyde	297	131	165	56
33	benzenepropanoic acid, methyl ester	138	47	91	66
S1	1,2,3-trimethoxybenzene	350	126	224	64
10:0 FAME	decanoic acid, methyl ester	280	155	125	45
34	3-methoxy benzoic acid, methyl ester	190	63	127	67
P5	1-(4-methoxy phenyl) ethanone	176	80	96	55
G3	4-ethenyl-1,2-dimethoxy-benzene	220	72	148	67
P6	4-methoxy-benzoic acid, methyl ester	301	125	176	58
36	1,2,4-trimethoxybenzene	621	235	385	62
S2	1,2,3-trimethoxy-5-methyl-benzene	100	39	60.6	61
40	3-methoxy-4-methylbenzoic acid, methyl ester	257	90	167	65
G4	3,4-dimethoxy-benzaldehyde	338	113	225	67
44	2-methoxy-benzenepropanoic acid, methyl ester	157	75	82	52
12:0 FAME	dodecanoic acid, methyl ester	257	134	123	48
G5	1-(3,4-dimethoxyphenyl) ethanone	220	85	135	61
G6	3,4-dimethoxy benzoic acid, methyl ester	675	123	552	82
S5	1-(3,4,5-trimethoxyphenyl)-ethanone	96	30	66.2	69
14:0 FAME	tetradecanoic acid, methyl ester	437	263	174	40
S6	3,4,5-trimethoxy-benzoic acid, methyl ester	258	49	209	81
FAME *iso* 15:0	12-methyl tetradecanoic acid, methyl ester	120	43	77	64
FAME *anteiso* 15:0	11-methyl tetradecanoic acid, methyl ester	263	51	211	80
16:1 ω 7 FAME	9-hexadecenoic acid, methyl ester	817	216	600	74
16:0 FAME	hexadecanoic acid, methyl ester	1380	779	601	44
18:1 ω 9 FAME	9-octadecenoic acid, methyl ester	229	93	136	59
18:0 FAME	octadecanoic acid, methyl ester	909	290	619	68
20:0 FAME	eicosanoic acid, methyl ester	55	20	35.5	64
22:0 FAME	docosanoic acid, methyl ester	61	15	45.7	75
24:0 FAME	tetracosanoic acid, methyl ester	81	23	58.1	72
26:0 FAME	hexacosanoic acid, methyl ester	39	10	29.1	74
	Total quantified aromatic contribution (ng L^{-1})	6115	2120	3995	65
	Total quantified lignin contribution (ng L^{-1})	3468	1302	2166	62
	Total quantified FAME contribution (ng L^{-1})	5030	2155	2875	57
	% of quantified aromatic compounds of total	23.8	19.1	4.6	19
	% of quantified lignin compounds of total	31.1	30.5	0.7	2
	% of quantified FAME of total	45.1	50.4	-5.3	-12
	Estimated lignin contribution (mg L^{-1})	0.105	0.0396	0.0658	62
	S/G ratio	0.41	0.40	0.01	3
	P/G ratio	0.46	0.87	-0.41	-89
	(Ac/Al)$_G$	2.00	1.09	0.90	45
	16:0 FAME / 26:0 FAME	35.0	74.9	-39.99	-114
	(*iso* 15:0 + *anteiso* 15:0) / 16:0	0.28	0.12	0.16	56

Table 3.11
Yields of TMAH products for White Clay Creek DOM and BDOM samples

Peak	Compound	Concentrations (ng L^{-1})			
		Inflow	Outflow	D ng	% Diff.
22	benzoic acid, methyl ester	517	207	310	60
8:0 FAME	octanoic acid, methyl ester	134	92.4	42.0	31
P3	1-ethenyl-4-methoxy-benzene	466	212	255	55
G1	1,2-dimethoxy-benzene	502	355	146	29
26	1,4-dimethoxy-benzene	924	270	654	71
31	2,5-dimethoxytoluene	166	132	34.0	20
	dimethoxytoluene	207	139	67.6	33
	3,4-dimethoxytoluene	400	282	118	30
P4	4-methoxy-benzaldehyde	508	123	386	76
33	benzenepropanoic acid, methyl ester	308	139	169	55
S1	1,2,3-trimethoxybenzene	451	145	306	68
10:0 FAME	decanoic acid, methyl ester	326	222	104	32
34	3-methoxy benzoic acid, methyl ester	273	141	132	48
P5	1-(4-methoxy phenyl) ethanone	317	150	167	53
G3	4-ethenyl-1,2-dimethoxy-benzene	515	154	361	70
P6	4-methoxy-benzoic acid, methyl ester	429	176	253	59
36	1,2,4-trimethoxybenzene	750	197	554	74
S2	1,2,3-trimethoxy-5-methyl-benzene	400	68.0	332	83
40	3-methoxy-4-methylbenzoic acid, methyl ester	451	243	208	46
G4	3,4-dimethoxy-benzaldehyde	411	195	217	53
44	2-methoxy-benzenepropanoic acid, methyl ester	301	202	99.8	33
12:0 FAME	dodecanoic acid, methyl ester	398	167	231	58
G5	1-(3,4-dimethoxyphenyl) ethanone	105	63.3	41.4	40
G6	3,4-dimethoxy benzoic acid, methyl ester	304	67.4	236	78
S5	1-(3,4,5-trimethoxyphenyl)-ethanone	55	22.6	32.6	59
14:0 FAME	tetradecanoic acid, methyl ester	334	191	144	43
S6	3,4,5-trimethoxy-benzoic acid, methyl ester	143	15.3	127	89
FAME *iso* 15:0	12-methyl tetradecanoic acid, methyl ester	240	70.6	169	71
FAME *anteiso* 15:0	11-methyl tetradecanoic acid, methyl ester	570	215	355	62
16:1 ω 7 FAME	9-hexadecenoic acid, methyl ester	2710	356	2354	87
16:0 FAME	hexadecanoic acid, methyl ester	999	419	580	58
18:1 ω 9 FAME	9-octadecenoic acid, methyl ester	1220	237	983	81
18:0 FAME	octadecanoic acid, methyl ester	206	182	23.9	12
20:0 FAME	eicosanoic acid, methyl ester	38	25.7	12.4	32
22:0 FAME	docosanoic acid, methyl ester	35	18.4	16.4	47
24:0 FAME	tetracosanoic acid, methyl ester	43	22.3	20.4	48
26:0 FAME	hexacosanoic acid, methyl ester	24	—	6.54	~100
	Total quantified aromatic contribution (ng L^{-1})	8903	3697	5206	58
	Total quantified lignin contribution (ng L^{-1})	4606	1766	2839	62
	Total quantified FAME contribution (ng L^{-1})	7253	2218	5034	69
	% of quantified aromatic compounds of total	26.6	32.6	-6.04	-23
	% of quantified lignin compounds of total	28.5	29.9	-1.35	-5
	% of quantified FAME of total	44.9	37.5	7.39	16
	Estimated lignin contribution (mg L^{-1})	0.140	0.0537	0.0863	62
	S/G ratio	0.46	0.22	0.237	52
	P/G ratio	0.75	0.60	0.154	20
	(Ac/Al)$_G$	0.74	0.35	0.392	53
	16:0 FAME / 26:0 FAME	41.7	—	—	—
	(iso 15:0 + anteiso 15:0) / 16:0	0.81	0.68	0.13	16

subtracting the yield of lignin-derived, aromatic products from the total quantified aromatic contribution (only the non-lignin aromatic products are represented) and dividing by the total summed yield of all TMAH products. The percent of lignin-derived products and FAMEs relative to total yield is calculated by dividing the total product yield for lignin-derived products by the total yield of all TMAH products.

For the Rio Tempisquito DOM and RDOM samples, there was a 57-65% decrease in total product yield for both lignin-derived and non-lignin-derived aromatic products and FAMEs (Table 3.10). This indicates that more than half of the initial concentration, as the sum of the precursors for these products, in the stream water was degraded during passage through the bioreactors. The precursors for both the lignin-derived and non-lignin-derived aromatic products were more effectively degraded within the bioreactors than those for FAMEs. This suggests that lignin residues and carbohydrates, proteins, or tannins contribute more to the molecular composition of BDOM than fatty acids in this stream ecosystem. In both the DOM and RDOM, FAMEs contributed more to the total product yield than either lignin-derived or non-lignin-derived aromatic products. This does not mean that fatty acids contributed more to the DOM or RDOM than lignin or the sum of carbohydrates, proteins, or tannins since we expect the yield of products from these latter compounds is substantially less than the yield from fatty acids, although this needs to be confirmed.

Based on TMAH yield of products found for pure degraded lignin (~3.3%; Frazier et al. 2003) one can estimate the concentration of lignin residues present in the stream water (Table 3.10) and calculate that lignin residues may have represented up to 16% of the stream water DOC and 7% of the DOC after degradation (calculated by dividing the concentration of lignin residues by the total DOC). The concentration of lignin residues in the stream water therefore decreased by 62% upon degradation during passage through the bioreactors (derived from the information in Tables 3.8 and 3.9). These results suggest that lignin residues present on DOM can be an important source of energy for heterotrophic bacteria in stream ecosystems.

Ratios of lignin-derived products such as the syringyl to guaiacyl (S/G, or the sum of all products having a source from syringyl structures in lignin to those from guaiacyl structures) or the p-hydroxyphenyl to guaiacyl (P/G) have been used to indicate the nature of the lignin sources to DOM (Ertel, Hedges, and Perdue 1984). Syringyl compounds result from angiosperms or flowering plants such as deciduous trees, guaiacyl compounds result from angiosperms and gymnosperms or plants that produce cones such as conifers and cycads, and p-hydroxyphenyl compounds result from non-woody vegetation. Therefore, the relative amounts of these compounds to each other can indicate the predominant lignin inputs from these different forms of vegetation. The S/G and P/G ratios indicate that the lignin originated predominantly from non-woody gymnosperms (Table 3.10). However, the surrounding vegetation is predominantly woody and non-woody angiosperms. Preferential degradation of specific lignin biomolecules may obscure the results for such calculations because certain components will reside longer in the DOM and complicate such interpretations. Also, these ratios were developed using CuO oxidation and, therefore, they may not directly apply to TMAH thermochemolysis. Hatcher and Minard (1996) found that the acid aldehyde ratios correlate well with CuO, but TMAH yielded higher values than CuO. It is likely that relationships could be developed for these ratios as well, but this has yet to be investigated.

These ratios are used here more to illustrate changes in DOM molecular composition during biodegradation. Bacteria within these bioreactors appear to have no preference between guaiacyl and syringyl forms of lignin as indicated by the consistent S/G ratio in the DOM and RDOM (Table 3.10). The P/G ratio increases upon degradation by the bioreactors (Table 3.10). However, p-hydroxyl phenyl compounds have been found to be products from other sources such as proteins (Ertel, Hedges, and Perdue 1984), but we do not expect proteins to contribute more to DOM after biodegradation. The selective utilization of amino acids in stream water has been demonstrated (Volk, Volk, and Kaplan 1997) and Ellis et al. (1999) found that amino acids provided the most favorable growth conditions for biofilms, followed by carbohydrates, and then humics. Therefore, p-hydroxyphenyl compounds in the RDOM may have origins other than proteins and may be predominantly lignin-derived. This interpretation suggests that the increase in the P/G ratio between the DOM and RDOM samples results from a preference of bacteria for guaiacyl forms of lignin over the p-hydroxyphenyl.

The acid to aldehyde ratio (($Ad/Al)_G$) is used to reflect the degree of degradation for lignin as microbes oxidize the α-carbons of aldehydes to produce acids. Therefore, as degradation proceeds, the $(Ad/Al)_G$ is anticipated to increase and this has been demonstrated with TMAH thermochemolysis for terrestrially degraded lignin (Hatcher et al. 1995). The $(Ad/Al)_G$ was also applied to the analysis of the state of oxidation of lignin in DOM using CuO (Ertel, Hedges, and Perdue 1984). In the Rio Tempisquito, the $(Ad/Al)_G$ decreases substantially between the DOM and RDOM samples (Table 3.10). In these bioreactors, bacteria preferred the more oxidized lignin to that which was less degraded, thus producing more of the aldehyde monomer than the acid monomer upon TMAH thermochemolysis. This suggests that, in stream ecosystems, the $(Ad/Al)_G$ may not be a useful indicator of the degree of degradation of lignin residues in DOM since one would have anticipated lignin residues in the RDOM to be more oxidized than in the DOM.

The 16:0/26:0 ratio for fatty acids is often used as an indicator of the relative microbial contributions to terrestrial contributions for NOM (Napolitano 1999). The ((*iso* 15:0 + *anteiso* 15:0)/16:0) ratios for fatty acids have been used as an indicator of bacterial contribution relative to contributions from eukaryotic organisms (Meyers and Ishiwatari 1993; Napolitano 1999). For this study these ratios were more useful as indicators of changes in organic matter composition. The 16:0/26:0 increased in the RDOM relative to the DOM, indicating that the organic matter was enriched with lower molecular weight (possibly microbially derived) fatty acids, considered more metabolically labile than the higher molecular weight fatty acids of higher plants. This suggests that there were bacterial contributions of fatty acids to DOM by passage through the bioreactors. The *iso*- and *anteiso*- 15:0 FAMEs are thought to be characteristic of gram-positive bacteria (Moll and Summers 1999). Both *iso*- and *anteiso*- 15:0 FAMEs were degraded more effectively within the bioreactors than the *n*-FAMEs as judged by the decreasing ratios (Table 3.10). The results demonstrate that the fatty acid composition of DOM can be altered upon bacterial degradation, though the distribution from fatty acids was, in general, qualitatively similar between the DOM and RDOM samples (Figure 3.7). Bacteria likely affect the fatty acid composition of DOM by both selectively degrading certain components of DOM while contributing others.

The TMAH product yield from the stream water BDOM (ng L^{-1}) can be measured as the difference between the yield of a product from the RDOM compared to the yield of a product from the DOM sample. Decreased yields are observed for all TMAH thermochemolysis products in the RDOM relative to the DOM. This indicates the precursors for all the observed products were degraded during passage through the bioreactors. The 16:1 ω7, 16:0, and 18:0 FAMEs were the most predominant FAMEs in the DOM sample and were also the most effectively degraded. The unsaturated fatty acids were degraded to a greater relative degree than the 16:0 FAME, which is expected since unsaturated fatty acids are more metabolically labile that the *n*-fatty acids. For the aromatic products, benzoic acid methyl ester (multiple precursors), 1,2,4-trimethoxy benzene (carbohydrate, possibly other sources), and 3,4-dimethoxy benzoic acid methyl ester (lignin) were the predominant products in the DOM and their precursors also showed the greatest degree of degradation. The precursors for other lignin- and carbohydrate-derived products were also effectively

47

degraded within the bioreactors. This suggests that both lignin residues and carbohydrates can represent an equivalent fraction of the BDOM. In general, the relative yield of one aromatic product to the next was similar between the DOM and BDOM. This suggests, assuming product yields from the precursors for the aromatic products were similar, that the most abundant compounds in DOM also contributed the most to the BDOM.

Quantitative Comparison of TMAH Products From the White Clay Creek DOM and RDOM

The yield of aromatic products from both lignin and non-lignin precursors and FAMEs decreased substantially between the DOM and RDOM samples (Table 3.11). The relative decreases in yield were similar for the lignin and non-lignin derived aromatic summed product yields, whereas the relative decrease in the summed yield for FAMEs was greater than that of aromatic products (Table 3.11). This suggests that fatty acids were more easily degradable than the precursors for the aromatic products (e.g. lignin, tannins, and carbohydrates) at the White Clay Creek site than those precursors contributing to the yield of aromatic products.

The estimated lignin contribution to DOM decreased substantially by passage through the bioreactors (Table 3.11). We estimated that lignin residues represent 3.7% and 2.0% of the organic matter in the DOM and RDOM, respectively (Table 3.11). These results suggest that heterotrophic bacteria can effectively degrade lignin in stream DOM. However, lignin did not contribute substantially to the DOM or BDOM at this site. The differences in S/G, P/G, and $(Ad/Al)_G$ ratios suggest that there was selective degradation between lignin monomers at this site. Syringyl and p-hydroxyphenyl compounds were apparently selectively degraded relative to guaiacyl as judging by the changes in S/G and P/G ratio. However, the degradation of proteins, which can also yield p-hydroxyphenyl products, could also contribute to the decreased P/G. These ratios indicate that there was predominant input of lignin to DOM from non-woody gymnosperms, which contrasts with the surrounding vegetation. The $(Ad/Al)_G$ ratio again decreased between DOM and RDOM indicating the bacteria may prefer oxidized methoxy aromatic structures over the more reduced material and presumably less degraded lignin (Table 3.11).

The 16:0/26:0 and ((*iso* 15:0 + *anteiso* 15:0)/16:0) suggest a predominant bacterial contribution to the DOM and RDOM pools and that there was little input from fatty acids derived from higher plants (those longer than C_{24}) (Table 3.11). The ((*iso* 15:0 + *anteiso* 15:0)/16:0) relationship again decreased between DOM and RDOM indicating selective removal of terminally branched fatty acids.

A decrease in TMAH product yield was observed for all quantified products (Table 3.11). Several fatty acids contributed to the BDOM in this stream ecosystem. The 16:1ω7, 16:0, and 18:1ω9 FAMEs were the most predominant FAMEs in the DOM and were the most efficiently degraded by the heterotrophic bacteria, whereas the other low-molecular weight fatty acids were degraded to a lesser degree. These results are consistent with prior studies that found unsaturated fatty acids more metabolically labile than saturated fatty acids (Kawamura, Ishiwatari, and Ogura 1987; Meyers and Ishiwatari 1993; Meyers et al. 1984; Napolitano 1999). The higher-molecular weight fatty acids present in the DOM were likely terrestrially derived, were not degraded during passage through the bioreactors, and therefore did not contribute to the BDOM.

All precursors for the aromatic TMAH products were degraded during passage through the bioreactors (Table 3.11). The carbohydrate precursors for 1,4-dimethoxy benzene and 1,2,4-trimethoxy benzene were the most effectively degraded in the bioreactors compared to other TMAH products. However, other possible carbohydrate-derived products, the dimethoxy toluenes, were not effectively degraded. This suggests that there may be some selectivity in degradation between different carbohydrates.

Most lignin precursors were degraded upon passage through the bioreactors aside from 1-(3,4-dimethoxyphenyl) ethanone and 1-(3,4,5-trimethoxyphenyl)-ethanone. These results and those of the $(Ad/Al)_G$ suggest that the bacteria within these bioreactors were less adept at using the less oxidized lignin residues in this DOM compared to the more oxidized lignin residues (those that produce the aldehyde and acid TMAH products). Therefore, we conclude that lignin was selectively degraded here subject to type and functional group substitution. The degraded DOM was enriched in guaiacyl compounds relative to syringyl and more oxidized lignin was preferentially degraded over the more reduced forms. The products of undetermined sources were not as effectively degraded compared to other TMAH products and therefore contributed less to the BDOM. Again, the most abundant compounds in DOM also contributed the most to the BDOM.

Comparison of the Quantitative TMAH Thermochemolysis GC-MS Results Between the Rio Tempisquito and the White Clay Creek DOM and RDOM Samples

In terms of the bulk properties interpreted from the quantitative TMAH results, lower aromatic, lignin, and FAME product yields were observed in the RDOM samples compared to the DOM samples for both sites (Tables 3.10 and 3.11). This indicates a commonality at both sites in that carbohydrates, lignin, and fatty acids contributed to BDOM. Though the bacteria colonizing the bioreactors at the White Clay Creek site more effectively utilized fatty acids, relative to all aromatic products, compared to the Rio Tempisquito site.

Lignin contributed more to the DOM at the Rio Tempisquito site than at the White Clay Creek site and was more effectively utilized by bacteria at the Rio Tempisquito site compared to the White Clay Creek site (Tables 3.10 and 3.11). There is evidence from both sites that lignin is selectively degraded and that this selectivity can be unique to individual stream ecosystems. Syringyl compounds were preferentially degraded relative to guaiacyl compounds in White Clay Creek bioreactors whereas there was no preference in the Rio Tempisquito bioreactors. Guaiacyl compounds were preferentially degraded relative to p-hydroxyphenyl compounds in the Rio Tempisquito bioreactors whereas there was no preference in the White Clay Creek bioreactors (Tables 3.10 and 3.11). These results suggest that information regarding the terrestrial sources of lignin may be obscured upon bacterial degradation of lignin in stream ecosystems.

The water from both of the stream ecosystems showed a substantial decrease in the $(Ad/Al)_G$ ratio between the DOM and RDOM samples. This is contrary to what has been observed for terrestrially degraded lignin (Ertel and Hedges 1984; Ertel, Hedges, and Perdue 1984). In fact, both the Rio Tempisquito and White Clay Creek bioreactors utilized the more oxidized precursors for 3,4-dimethoxy benzoic acid methyl ester (G6) and 3,4,5-trimethoxy benzoic acid methyl ester (S6) more effectively than the more reduced lignin precursors, those that produce aldehyde and ketone TMAH products (Tables 3.6 and 3.7). Also, the degradation of 3,4,5-trimethoxy benzoic acid by *Pseudomonas spp.* has been found only to occur in the presence of 3,4-dimethoxy benzoic acid and they were degraded at nearly equivalent rates, which was observed here (Kawakami 1980). It has been reported that bacteria easily rupture aromatic rings when a carboxyl group or a functionality easily oxidized to a carboxyl group is accompanied by a hydroxyl group in the *para* position (Kawakami 1980). The degradation of such lignin subunits, which likely produce 3,4-dimethoxy benzoic acid methyl ester upon treatment with TMAH, may occur more quickly compared to the lignin residues in DOM that have more reduced side-chains and that produce 3,4-diemthoxy benzaldehyde upon treatment with TMAH. Therefore, if side chain oxidation of the more reduced lignin residues were slower than ring rupture of the more oxidized lignin residues the $(Ad/Al)_G$ may be expected to decrease upon degradation by heterotrophic bacteria.

These results indicate that $(Ad/Al)_G$ may not accurately reflect the degree of degradation of lignin in natural waters as anticipated previously (Ertel, Hedges, and Perdue 1984). The acid/aldehyde ratios described by Ertel, Hedges, and Perdue (1984) relationships were based upon terrestrial sources of organic matter that had undergone terrestrial degradation, where fungi are the primary decomposers, and have not been verified for similar substances in natural waters, where bacteria are the primary decomposers. Also, the most abundant, TMAH amenable compounds in DOM were also the most effectively degraded during passage through the bioreactors. This suggests that the indigenous bacterial communities were adapted or selected to use the most abundant food source.

^{13}C TMAH Thermochemolysis GC-MS

The ^{13}C-TMAH thermochemolysis GC-MS procedure was used to further investigate the effects of bacterial degradation on White Clay Creek DOM molecular composition. The ^{13}C-TMAH procedure allows one to distinguish between compounds having methoxy functionalities prior to TMAH thermochemolysis and those that acquire methoxyl groups from the TMAH procedure. Noting the presence of 100% ^{13}C methylation in a compound by mass spectrometry is a clear determination that the precursor was not methylated prior to ^{13}C TMAH thermochemolysis. Hence by analysis of the products produced from the ^{13}C-TMAH procedure, we were able to observe molecules in White Clay Creek DOM that possessed natural methoxy groups and also how bacterial degradation affected these molecules. In the following discussion we use the term precursor to refer to the compound from which the ^{13}C-labeled TMAH product derives. Here the precursor is assumed to be of a similar structure to the product, differing only by the initial presence of a $-OCH_3$ or a $-OH$ functionality. However, we are aware that there can be other modifications during TMAH thermochemolysis than this from the original structure of molecules, as they were present in DOM (see Filley, Minard, and Hatcher 1999 for examples).

The molecular weight of the dimethoxy benzene TMAH products is 138 and this can be observed by mass spectrometry. Since these products contain two methoxy groups they can acquire up to two ^{13}C-methyl groups. If the precursors for the dimethoxy benzene products have 2 methoxy groups prior to ^{13}C-TMAH then they will not acquire any ^{13}C methyl groups and the molecular weight will remain 138. However, if the precursors for the dimethoxy benzene products have one or no methoxy groups prior to TMAH thermochemolysis then they will acquire one and two ^{13}C methyl groups and the molecular weight will be 139 and 140, respectively. By using equations 2 through 8, presented in the Methods section, the relative contribution from each of the precursors can be determined.

For the compounds in the following discussion, listed in Table 3.6, there were no original methoxy groups on their precursors and were, therefore, all initially phenolic, acidic, or ester linked and became methylated with ^{13}C methyl groups. The methyl ester group for all FAMEs and diacid methyl esters were ^{13}C-labeled as a result of the ^{13}C-TMAH procedure and were therefore either free acids or ester linked to molecules larger than a methyl group. The spectra for the furanose carbohydrate products were identical to unlabeled spectra except for 2-ethyl-3-methoxy-2-cyclopenten-1-one (PS7) where 100% of the product was enriched with one ^{13}C-label. This is expected since carbohydrates would not have preexisting methoxy groups. The dimethoxy-methyl benzene isomers (25, G2) and 3,4,5-trimethoxy toluene (S2) had no original methoxy groups prior to the reaction. These compounds can result from precursors such as carbohydrates and tannins so the lack of preexisting methoxy groups is consistent with these sources. Some of the products of uncertain or less specific origin, such as benzyl methyl ether (8); methoxy-methyl benzenes (9,10); methoxy benzene acetic acid methyl esters (31); and benzoic acid methyl ester (17) were also fully ^{13}C methylated. Benzene propanoic, propenoic, and their methoxy substituted methyl esters (30, 37) have been found to result from proteins (Mannino and Harvey 2000) so the lack of any original methoxy groups is consistent with these origins. Methoxy benzene (P1); 1-ethenyl-4-methoxy benzene (P3); 4-methoxy benzaldehyde (P4); and 4-methoxy benzoic acid methyl ester (P6) can result from proteins and lignin and the lack of original methoxy groups is consistent with these sources. 3-methoxy benzoic acid methyl ester (27) and 3-methoxy-4-methyl benzoic acid methyl ester (33) result from undetermined sources and were not methylated prior to treatment with TMAH. 1,1'-(1-methylidene) bis-4-methoxy benzene (39, both methoxy groups labeled) derives from bisphenol A.

The ^{13}C TMAH results for 1,2-, 1,4-, and 1,3-dimethoxy benzene are presented in Table 3.12. From the mass spectra we determined that the 1,2-dimethoxy benzene TMAH product (G1) from DOM derived from 32.2% 2-methoxy phenol and 67.8% catechol. There was a slight shift towards catechol in the RDOM relative to the DOM, however, considering the experimental variance this is likely insignificant. 1,2 dimethoxy benzene, presumably as catechol prior to methylation, has been found to be a TMAH product from degraded lignin (del Rio et al. 1998) and can also result from carbohydrates (Fabbri and Helleur 1999). We found in preliminary experiments that 1,2- and 1,4-dimethoxy benzene were produced in equivalent amounts from treatment of carbohydrates with TMAH (unpublished data). Also, one may expect bacteria to preferentially attack carbohydrates over lignin thus increasing the relative amount of 2-methoxy phenol (Ellis et al. 1999). Since the yield from both possible precursors for 1,2-dimethoxy benzene was consistent between the DOM and RDOM samples (insignificant apparent degradation), much of the yield of 1,2-dimethoxy benzene may result from lignin that was relatively refractory towards bacterial degradation. In addition, the catechol precursor could also result from tannins—also accounting for its refractory nature. Regardless, the precursors for this product did not contribute significantly to the BDOM in this stream.

Nearly all of the 1,4-dimethoxy benzene product from both the DOM and RDOM samples resulted from the 1,4-dihydroxy benzene precursor (Table 3.12). Since 1,4-dimethoxy benzene has been found to derive from carbohydrates, the absence of natural methoxy groups on the precursor for this compound was consistent with this source. The yield of 1,4-dimethoxy benzene decreased in the RDOM compared to the DOM indicating the precursor(s) for this compound, possibly carbohydrates, were effectively degraded by the bioreactors.

For the 1,3-dimethoxy benzene product, the majority of the product was derived from 3-methoxy phenol, whereas 3-hydroxy phenol accounted for less. In the RDOM, 3-hydroxy phenol accounted for the majority of the product (Table 3.12). This may indicate that bacterial communities can demethylate methoxy-substituted aromatic compounds. Carbohydrates did not produce detectable amounts of 1,3-dimethoxy benzene in our lab (unpublished data). Therefore, an argument for dilution by carbohydrate products is less likely. This result is significant for two reasons: bacteria are not typically considered efficient at demethylating methoxy-substituted aromatic compounds (e.g. lignin) and a source for 3-methoxy phenol in nature is undetermined.

Table 3.12
^{13}C-labeled TMAH thermochemolysis results for the WCC DOM, RDOM, and BDOM

Compound	Units*	DOM			RDOM			BDOM		
		2 ^{13}C†	1 ^{13}C‡	No ^{13}C§	2 ^{13}C†	1 ^{13}C‡	No ^{13}C§	2 ^{13}C†	1 ^{13}C‡	No ^{13}C§
1,2-dimethoxy benzene	%		32.2	67.8		28.7	71.3		3.5	-3.5
	ng L^{-1}		162	340		102	253		59.8	87.2
1,4-dimethoxy benzene	%		4.6	95.4		3.4	96.6		1.2	-1.2
	ng L^{-1}		42.5	881		9.15	260		33.4	622
1,3-dimethoxy benzene	%		54.8	45.2		45.4	54.6		9.4	-9.4
	ng L^{-1}		—	—		—	—		—	—
1,2,3-trimethoxy benzene	%	51.7	4.6	43.6	46.9	17.3	35.9	4.8	-12.7	7.7
	ng L^{-1}	233	20.7	197	68.0	25.1	52.1	165	-4.34	145
1,3,5-trimethoxy benzene	%	9.9	29.4	60.7	10.4	26.4	63.2	-0.5	3	-2.5
	ng L^{-1}	—	—	—	—	—	—	—	—	—
(G4)	%		93	7		94	6		-1	1
	ng L^{-1}		382	28.8		182	11.6		200	17.1
(G5)	%		85.7	14.3		85.6	14.4		0.1	-0.1
	ng L^{-1}		90.0	15.0		53.9	9.07		36.1	5.94
(G6)	%		57	43		41.1	58.9		15.9	-15.9
	ng L^{-1}		173	131		27.5	39.5		146	91.3

* The quantity of the precursor was calculated as percent of total TMAH product or as yield L^{-1}.
† The amount of precursor that possessed 2 natural methoxy groups.
‡ The amount of precursor that possessed 1 natural methoxy group.
§ The amount of precursor that possessed no natural methoxy groups.

The results for the dimethoxy benzene products also suggest that there was isomeric selectivity in the demethylation of different methoxy substituted, aromatic structures in DOM. The precursors for 1,2-dimethoxy benzene were not apparently demethylated whereas the precursors for the 1,3-dimethoxy benzene isomer were. Since lignin biodegradation by bacteria has been assessed using distinct, culturable species of bacteria to degrade isolated lignin and model lignin compounds, which may not accurately represent the native microbial communities or native lignin in DOM, the efficiency of bacterial demethylation could have been overlooked to this point.

The TMAH product 1,2,3-trimethoxy benzene (S1) has three precursors: 2,5-dimethoxy phenol (presumably, from lignin), 1-hydroxy-5-methoxy phenol (presumably from degraded lignin), and 1,2,3-trihydroxy benzene (phloroglucinol, from degraded lignin, tannins, and carbohydrates) (Table 3.12). The dimethoxy phenol precursor was apparently demethylated or degraded within the bioreactors as suggested by the decreased yield from the RDOM sample.

The demethylation of the dimethoxy precursor may account for the increased concentration of the precursor with a single natural methoxy (1-hydroxy-5-methoxy phenol) group in the DOM sample. The yield from the 1,2,3-trihydroxy benzene precursor also decreased substantially in the RDOM relative to the DOM, indicating carbohydrates and degraded lignin contribute to the BDOM.

1,3,5-trimethoxy benzene has three possible precursors including dimethoxy phenol, hydroxy-methoxy phenol, and 3,5-dihydroxy phenol (Table 3.12). Both dimethoxy phenol and hydroxy-methoxy phenol were from undetermined sources and 3,5-dihydroxy phenol could be a product from carbohydrates and hydrolyzable-tannins. There were no substantial differences in the relative contributions to the products between the three precursors. Since there was demethylation of the precursors for 1,2,3-trimethoxy benzene, but not for the precursors for 1,3,5-trimethoxy benzene, there may also be preferential demethylation between these isomers as observed for the dimethoxy benzene products.

Lignin residues were best represented here by guaiacyl ^{13}C-TMAH products since the low signal to noise for the syringyl products precluded an acceptable level of separation from their co-eluting compounds and since p-hydroxyphenyl compounds may result from sources other than lignin (Table 3.12). 3,4-dimethoxy benzaldehyde (G4) has two possible precursors: 3,4-dihydroxy benzaldehyde (from degraded lignin) and 3-methoxy-4-hydroxy benzaldehyde (from more intact lignin). 3-methoxy-4-hydroxy benzaldehyde was almost exclusively the precursor for this TMAH product and there was no apparent demethylation of this precursor within the bioreactors. Here both precursors contributed equally to the BDOM in terms of relative amounts. Therefore, there was no selectivity in degradation between the two precursors. The TMAH product 1-(3,4-dimethoxyphenyl) ethanone (G5) also had two precursors: 1-(3-methoxy-4-hydroxyphenyl) ethanone (from intact lignin) and 1-(3,4-dihydroxyphenyl) ethanone (from degraded lignin). Again, the product yield of 1-(3,4-dimethoxyphenyl) ethanone was almost exclusively derived from that of intact lignin with only 14.3% of the demethylated precursor contributing to the product yield. There was also no demethylation of 1-(3-methoxy-4-hydroxyphenyl) ethanone and little degradation of either precursor during passage through the bioreactors. This indicates the lignin residues responsible for these TMAH products were not demethylated during degradation within bioreactors.

The two precursors that resulted in 3,4-dimethoxy benzoic acid methyl ester (G6) were 3-methoxy-4-hydroxy benzoic acid (from lignin) and 3,4-dihydroxy benzoic acid (from degraded lignin) (Table 3.10). Initially, 3-methoxy-4-hydroxy benzoic acid was the dominant precursor relative to 3,4-dihydroxy benzoic acid, but after biodegradation within bioreactors 3,4-dihydroxy benzoic acid was the dominant precursor. There was also a substantial decrease in product yield in the RDOM for both products. Therefore, the lignin residues responsible for these products may significantly contribute to the BDOM. This again shows that bacteria can demethylate lignin, a result that extends prior observations of lignin demethylation by bacterial species exposed to model lignin compounds. Some species of bacteria have been found to demethylate model lignin compounds, but this often requires substantial time scales (Kawakami 1980). This study in combination with the quantitative TMAH GC-MS also suggests that the more oxidized lignin residues in DOM were more susceptible to bacterial attack than the corresponding more reduced forms. These results support the conclusion that the degradation of lignin residues present in DOM by in-stream microbial communities can be selective.

OBJECTIVE 1C. THE INFLUENCE OF TREATMENT ON NOM AND BOM

Two data sets provided information on the impact of treatment processes on NOM and BOM. The first, initially presented in Tables 3.1 - 3.3 shows the influence of conventional treatment from the Indiana American water treatment plant on water from the White River. White River DOC concentrations in the raw water were reduced by 27 to 39% by settling and the DOC in settled water was reduced another 4 to 9% by filtering. Humic substances also declined during treatment, with reductions of 35 to 54% by settling, and another 0 to 15% by filtering. Lastly, the complex carbohydrates declined 54 to 70% upon settling, and then declined another 8 to 41% during filtering. Clearly, the greatest absolute reductions in DOC, humic-DOC and carbohydrates occurred as a result of settling.

The second example, presented below, describes the application of NMR to the Betasso pilot plant. The NMR method was utilized to examine samples associated the Betasso pilot plant that included ozonation and rapid sand filters (see Figure 2.1, p. 7). We examined numerous questions regarding the effects of sand filtering versus bioreactors, the effects of ozonation on the efficacy of sand filters and bioreactors, and the effects of temperature on DOM removal. Samples were collected from various points along the treatment path and dried according to our stated procedure.

In Figure 3.9 are ^{13}C CPMAS NMR spectra for a raw inflow water from Betasso WTP (sample 193), a water sample collected after a pre-ozonation procedure, and a set of samples collected downstream of sand filters placed after each of a pre-ozonation and intermediate ozonation process. All spectra show an intense signal for bicarbonate at 167 ppm and siloxanes at near 0 ppm. These signals were not quantitative and were not due to DOM in the samples, so their intensities are irrelevant to the following discussion. Comparing spectra 193 and 194 shows the effect of preozonation. It is clear that the main difference between these two spectra is a relative enhancement in carboxyl/amide functionalities (175 ppm) associated with oxidation of the DOM by ozone used in this process. Other signals in the spectra are not clearly affected by the process. This could be because the signals are broad and subtle changes are not easily discerned by the NMR technique. Passing the pre-ozonated water through a sand filter (sample 190) had no discernible relative effect on the carboxyl functionality. Intermediate ozonation followed by sand filtering may show a slightly less intense signal for carboxyl/amide functional groups compared with pre-ozonation and sand filtering, but the NMR spectra were noisy enough to preclude us from making this conclusion with much confidence. In fact all the spectra in Figure 3.9 are essentially the same. While the NMR approach is not overly sensitive to subtle changes within a treatment facility, there was a marked difference between the DOM from the Barker Lake water and that from Rio Tempisquito that appeared to be much more aromatic (100-160 ppm).

Figure 3.10 shows the NMR data for other samples collected from the Betasso WTP facility. Again, lack of significant differences limited the usefulness of this technique for identifying subtle changes in the nature of DOM associated with sand filtration as compared with bioreactor transformations. Sample 189 is DOM from pre-ozonated water passed through a bioreactor at ambient temperature.

TMAH Thermochemolysis of DOM Samples From Betasso WTP Facilities

The TMAH thermochemolysis GC-MS traces for the Betasso WTP samples are presented in several figures in sections below describing the water treatment activity involved at each stage of sample collection. The data discussed below focuses mainly on compounds whose quantitative measurements were readily available with the current inventory of standards.

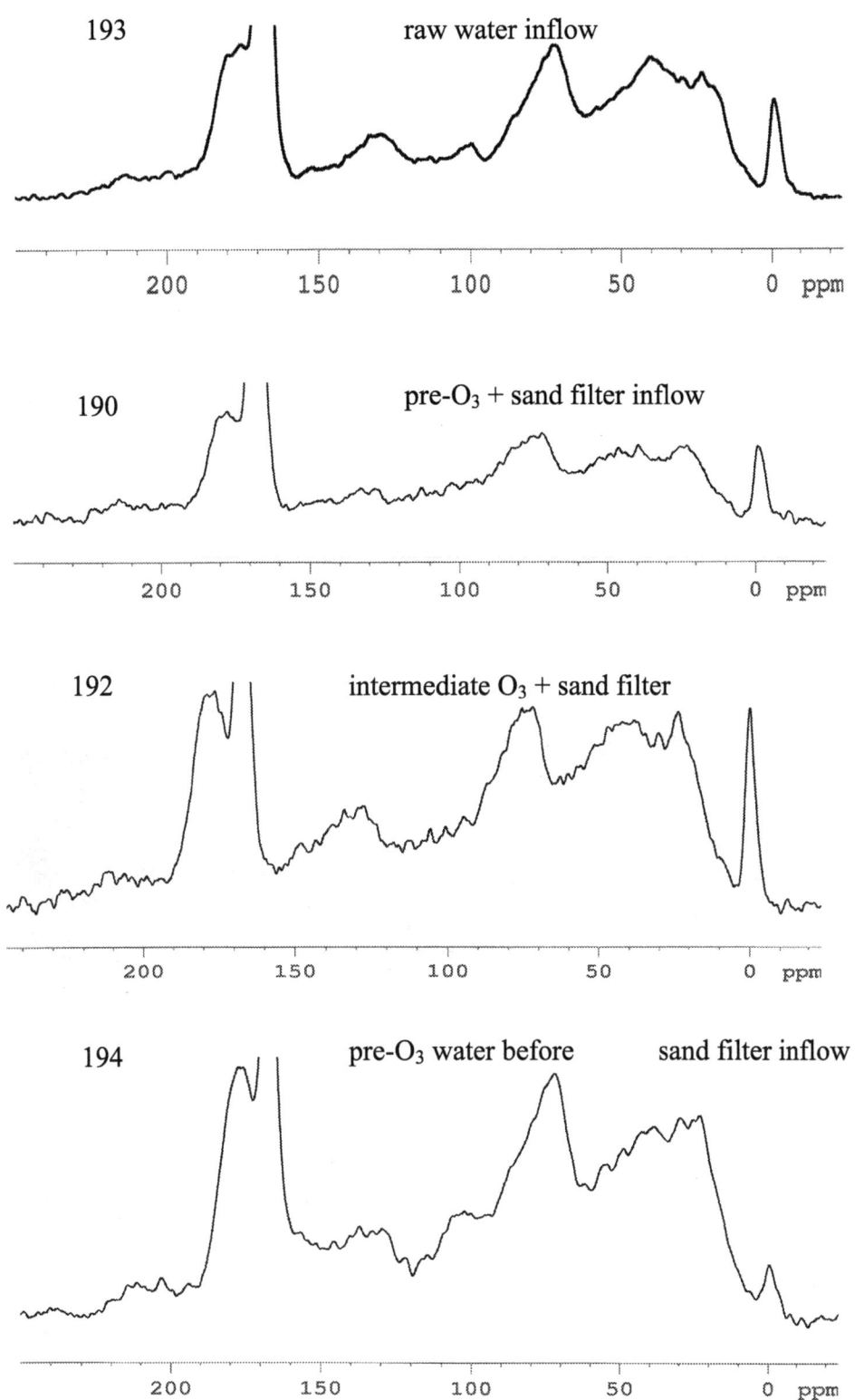

Figure 3.9 Solid-state ^{13}C NMR spectra of DOM from water used to feed RSF at the Betasso WTP

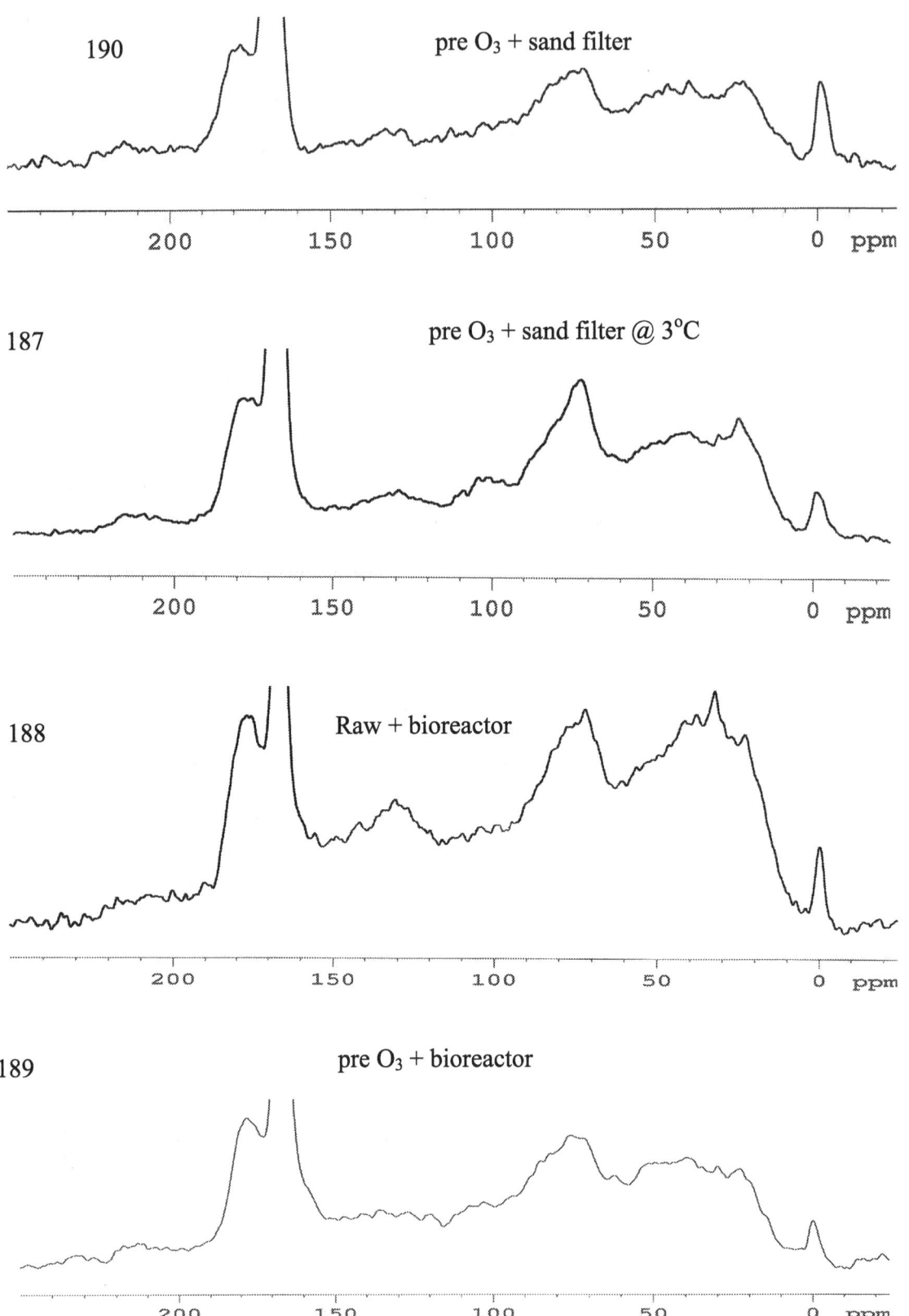

Figure 3.10 Solid-state ^{13}C NMR spectra of DOM in effluent waters from RSF and bioreactors at the Betasso WTP

Table 3.13
**Concentrations of products in samples of raw water (193) and
pre-ozonated water (194) from the Betasso WTP**

Compound	Concentrations (ng L^{-1})		
	Raw water	After preozonation	% Diff.
(P1) benzene, methoxy-	18.7	25.2	34.5
(P2) benzene, 1-methoxy-4-methyl-	19.5	27.8	42.4
(G1) benzene, 1,2-dimethoxy-	47.5	53.9	13.6
(G2) 3,4-dimethoxytoluene	17.9	27.0	51.1
(S1) 1,2,3-trimethoxybenzene	33.8	41.9	24.0
(P6) benzoic acid, 4-methoxy-, methyl ester	72.7	84.5	16.2
(S2) benzene, 1,2,3-trimethoxy-5-methyl-	18.3	27.0	47.3
(G4) benzaldehyde, 3,4-dimethoxy-	19.4	19.3	-0.7
(G5) ethanone, 1-(3,4-dimethoxyphenyl)	47.5	60.4	27.3
(G6) benzoic acid, 3,4-dimethoxy-, methyl ester	88.5	83.9	-5.2
(S6) benzoic acid, 3,4,5-trimethoxy-, methyl ester	32.0	40.1	25.3
octanoic acid, methyl ester	45.4	65.0	43.0
undecanoic acid, methyl ester	46.7	66.3	41.9
dodecanoic acid, methyl ester	22.5	53.7	138.6
methyl tetradecanoate	43.9	76.5	74.5
9-hexadecenoic acid, methyl ester, (Z)-	31.2	219.8	603.4
hexadecanoic acid, methyl ester	35.4	142.0	301.3
9-octadecenoic acid (Z)-, methyl ester	38.2	42.1	10.2
octadecanoic acid, methyl ester	10.5	33.4	218.0
eicosanoic acid, methyl ester	4.1	10.5	159.3
docosanoic acid, methyl ester	3.1	10.9	248.3

Comparison of Pre-ozonation of Intake Raw Water (Samples 193 and 194)

The TMAH thermochemolysis data for raw water (sample 193) contained three major types of compounds that can be quantified readily (Table 3.13, Figure 3.11), fatty acids with an even number of C atoms, lignin-derived products and alkanes. These products were common in many of the DOM samples examined previously and discussed above. A hump in the spectrum at a retention time of around 900 s was due to an unresolved complex mixture commonly observed in petroleum or biodegraded petroleum. The spectrum from the pre-ozonated sample (194) did not contain this hump. Instead, it displayed very intense peaks from 16:0 and 16:1 ω FAMEs. Quantitative analysis (Table 3.13) showed that there was a significant increase in concentrations of FAMEs after ozonation treatment. The concentration of 16:1 ω FAME was increased by 600%. This is consistent with results obtained from other treatment plants employing ozonation (Frazier 2001). It is likely that the ozonation process physically disrupted microorganisms, thereby

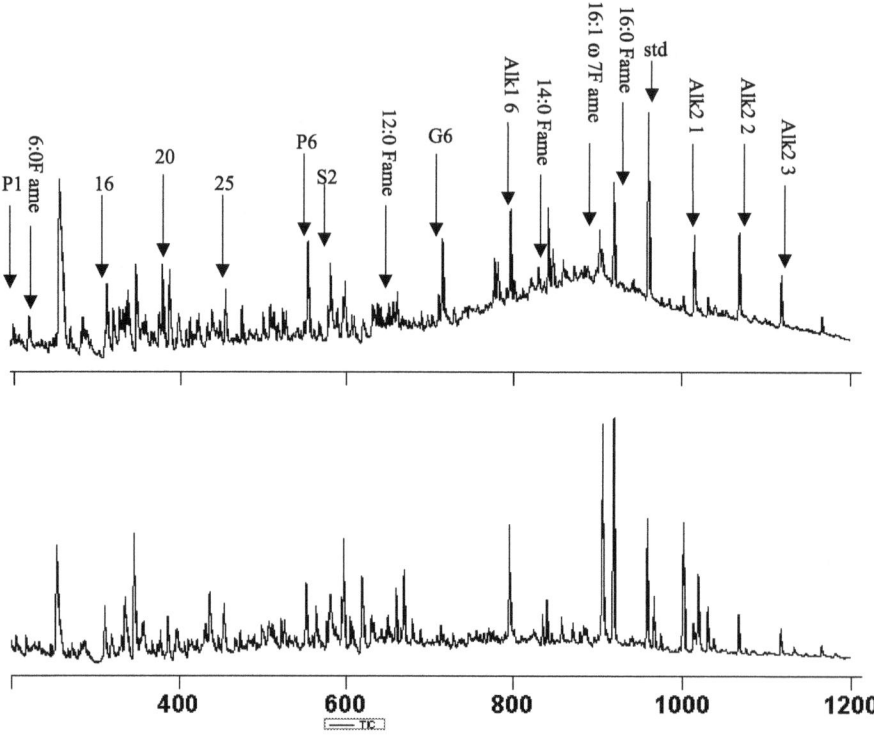

Figure 3.11 Chromatograms for TMAH products from samples 193 (upper) and 194 (lower)

increasing the contributions of released fatty acids into the DOM. The concentrations of lignin derived compounds also increased; however the changes were smaller than those from FAMEs. This result appears unexpected because ozone is known to be a strong oxidizating reagent for lignin polymers and was expected to selectively attack aromatic structures like lignin. It is possible that the oxidation reaction only depolymerizes lignin polymers contributing to the increase of lignin monomer concentrations. Only semi-quantitative analyses and comparisons were performed for alkanes, using areas obtained from GC chromatograms and assuming a unit response factor. There was little change between areas of alkanes from raw water and treated water samples, indicating that the ozonation was ineffective in breaking down alkanes, as expected.

Effects of Ozonation on Rapid Sand Filtration

The influence of ozonation treatment on removal of NOM was investigated by comparing water passed through a sand filter (sample 191) with water first pre-ozonated (sample 190) or treated with intermediate ozonation (sample 194) prior to sand filtration. The TMAH thermochemolysis chromatograms are shown in Figure 3.12. Quantitative data are presented in Table 3.14. The untreated water sample passed through a sand filter showed higher concentrations of lignin and fatty acids compared to other samples. The concentrations of compounds such as P1, P2, octanoic acid methyl ester and decanoic acid methyl ester were around 200% higher in the untreated sample with sand filter only. It is clear that that ozonation combined with sand filtration (pre-ozonation and intermediate ozonation) made the removal of compounds in DOM more efficient. No notable differences were observed beteween samples that had been treated by pre- and intermediate ozonation, followed by sand filtration.

58

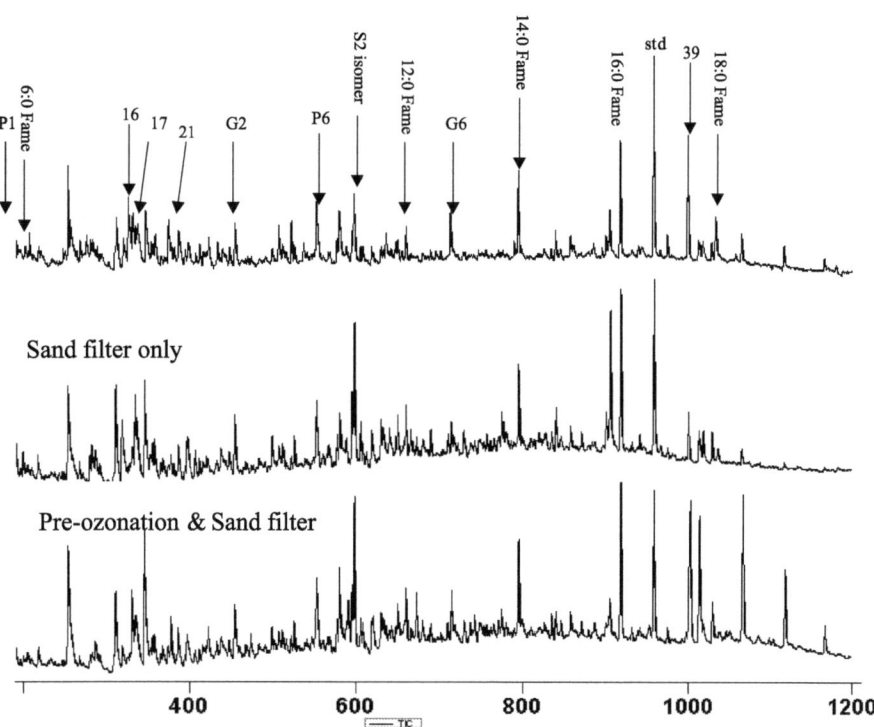

Figure 3.12 TMAH chromatograms of water samples from the Betasso WTP

OBJECTIVE 2A. SPECIES COMPOSITION OF HETEROTROPHIC BACTERIAL COMMUNITIES IN SOURCE WATERS FROM DIFFERENT WATERSHEDS

Communities that developed in the bioreactors were uniquely different among watersheds. Reproducibility between replicate bioreactors was high, as demonstrated by the samples from White Clay Creek in April (90-100%) (Figure 3.13). In temperate systems, there was a greater shift in community patterns associated with season than in the tropical system. In White Clay Creek (WCC), the November t-RFLP patterns were 30% similar to the patterns found in April. In contrast, the May and November sampling of the tropical system bioreactor from the Rio Tempisquito (RT) were 85% similar. Other bioreactor communities from temperate (Indiana American - IND) and semi-tropical (Hillsborough River - HB) systems were most similar to White Clay Creek (Figure 3.13). All of the bioreactors had roughly similar densities of bacterial cells, as determined by epifluorescent direct microscopic counts. Bioreactors colonized on the different water sources had densities that were estimated at 1.1×10^9 cells/gdw (gram dry weight) in the Betasso WTP bioreactors, 1.2×10^9 cells/gdw (White River), 2.1×10^9 cells/gdw (Hillsborough River), 1.4 to 3.6×10^9 cells/gdw (Rio Tempisquito), and 0.8 to 6.1×10^9 cells/gdw (White Clay Creek).

Similarity between some communities suggested that there was a cosmopolitan distribution of certain populations as inferred by common t-RFLP peaks. However, since t-RFLP analysis can group different bacterial species together, it was difficult to assess whether the ubiquitous t-RFLP peaks were due to common species distributed between systems or different species that had the same t-RFLP pattern. Future studies are needed to specifically relate common t-RFLP peaks to specific populations within each of the study systems.

Table 3.14
Concentrations of products from Betasso WTP samples

Compounds	Concentrations (ng L^{-1})		
	Preozonation & sand filter	Sand filter	Intermediate ozonation & sand filter
(P1) benzene, methoxy-	17.0	49.5	15.1
(P2) benzene, 1-methoxy-4-methyl-	19.6	53.8	16.9
(G1) benzene, 1,2-dimethoxy-	35.3	107.6	35.5
(G2) 3,4-dimethoxytoluene	35.2	50.8	16.9
(S1) 1,2,3-trimethoxybenzene	28.3	74.2	27.3
(P6) benzoic acid, 4-methoxy-, methyl ester	57.0	171.6	57.9
(S2) benzene, 1,2,3-trimethoxy-5-methyl-	25.1	47.3	16.7
(G4) benzaldehyde, 3,4-dimethoxy-	11.8	31.5	14.4
(G5) ethanone, 1-(3,4-dimethoxyphenyl)	44.5	124.8	40.9
(G6) benzoic acid, 3,4-dimethoxy-, methyl ester	62.1	161.9	69.1
(S6) benzoic acid, 3,4,5-trimethoxy-, methyl ester	27.8	79.9	26.6
octanoic acid, methyl ester	43.3	127.2	37.5
undecanoic acid, methyl ester	45.6	124.8	41.0
dodecanoic acid, methyl ester	32.4	38.5	34.1
methyl tetradecanoate	43.9	73.7	44.5
9-hexadecenoic acid, methyl ester, (Z)-	91.3	85.2	69.9
hexadecanoic acid, methyl ester	59.3	62.1	29.6
9-octadecenoic acid (Z)-, methyl ester	27.8	68.8	21.5
octadecanoic acid, methyl ester	15.5	19.4	11.6
eicosanoic acid, methyl ester	5.0	9.9	4.6
heneicosanoic acid, methyl ester	5.2	8.3	4.5

The t-RFLP method has similar sensitivity compared to other PCR based community analysis methods in that additions of plasmid DNA to DNA extracted from a natural sample with plasmid DNA showed that the t-RFLP method can detect populations comprising 0.1 to 1% of the bacterial community (Dunbar, Ticknor, and Kuske 2000). In general, however, tRFLP profiles do not reflect cell counts of natural microbial communities. There are complications at both the community level and the methodological level that would confound such a determination. At the community level, the composition of the microbial community is variable. Therefore, the genome size per cell and, similarly, the copies of the 16S rRNA gene per cell are variable. Since PCR amplifies genes and we do not know how many copies of the 16S rRNA gene there are per cell in a mixed community, it is hard to estimate the number of cells that can be detected. In this work, we used about 30 ng of community DNA per PCR reaction. Since we have clone libraries, we have an idea of the groups represented in the extracted DNA and can assume an average genome size of

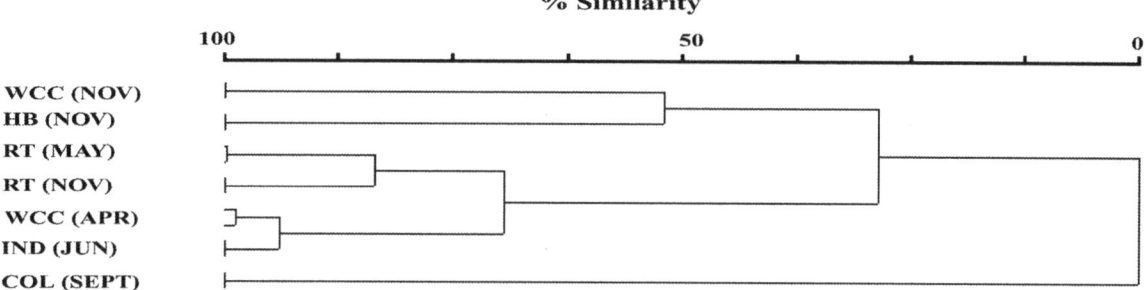

Figure 3.13 Regional and seasonal changes in t-RFLP patterns from bioreactor communities. WCC (NOV) = White Clay Creek, PA (November); HB (NOV) = Hillsborough River, FL (November); RT (MAY) = Rio Tempisquito, Costa Rica (MAY); RT (NOV) = Rio Tempisquito, Costa Rica (November); WCC (APR) = White Clay Creek, PA (April); IND (JUN) = White River, Indiana (June); COL (SEPT) = Betasso WTP, CO (September).

3.6×10^6 basepairs/cell. The number of cells in the extractable DNA pool prior to PCR amplification is 1.2×10^7 cells extracted from 1 gram of sediment (Fogel et al. 1999). At the methodological level, there are biases due to PCR and differential extraction of different microbial groups that alter the apparent distribution of bacterial types in the amplifiable PCR pool as compared to the distribution of bacterial types in nature. In our clone libraries, we have good representation on organisms that are known to be hard to extract DNA from such as Cyanobacteria, Gram-positive organisms, and Planctomycete-like organisms. This suggests an efficient extraction procedure. However, in a natural sample it is impossible to assess the bias due to PCR because we do not know the sequences of the organisms that are in the community.

Further investigation of the bacterial community from the Rio Tempisquito used a phylogenetic analysis of 16S rDNA clones recovered from the bioreactors, epilithon, and sediments of Rio Tempisquito to evaluate diversity. This revealed a broad phylogenetic distribution (Figure 3.14 -3.16). Samples analyzed were all collected in May 1999. Bioreactor and sediment communities had clones representative of the Beta and Gamma Proteobacteria (Figure 3.14). Phylotypes involved in nitrification were also present in the bioreactor and sediment communities. Sediment clone, Pool clone 54I, was most closely related to *Nitrosococcus sp* C-113 whereas the bioreactor clone (bioreactor clone 142-2) was most closely related to *Nitrosomonas* europea (Figure 3.14). Clones from the replicate bioreactors also clustered closely together within the Cytophaga/Flexibacteria/Bacteroides (Figure 3.15). One clone from the epilithon was closely related to *Sphingomonas* (Figure 3.15). Three clones from the Bioreactor 153 were closely related to *Planctomyces* (Figure 3.16). Once we determined the phylogenetic placement of the clones from Rio Tempisquito, we identified the clones in tRFLP traces from bioreactors and sediments from Rio Tempisquito (Figure 3.17). *Cyanobacteria* (clone 70C) were found in both communities although at reduced levels in the bioreactor. This is not surprising since the bioreactors are incubated under darkened conditions. *Cytophaga* (clone 70A) were also found in both systems. *Cytophaga* are strictly aerobic or facultatively anaerobic chemoorganoheterotrophs, capable of metabolizing a broad range of DOC compounds (Reichenbach 1992). Bacteria involved in nitrogen cycling were present in the sediment samples but not in the bioreactor. A *Planctomycete* (clone 153E), which is another aerobic bacteria capable of metabolizing a broad range of carbon compounds, was identified in the sediment samples.

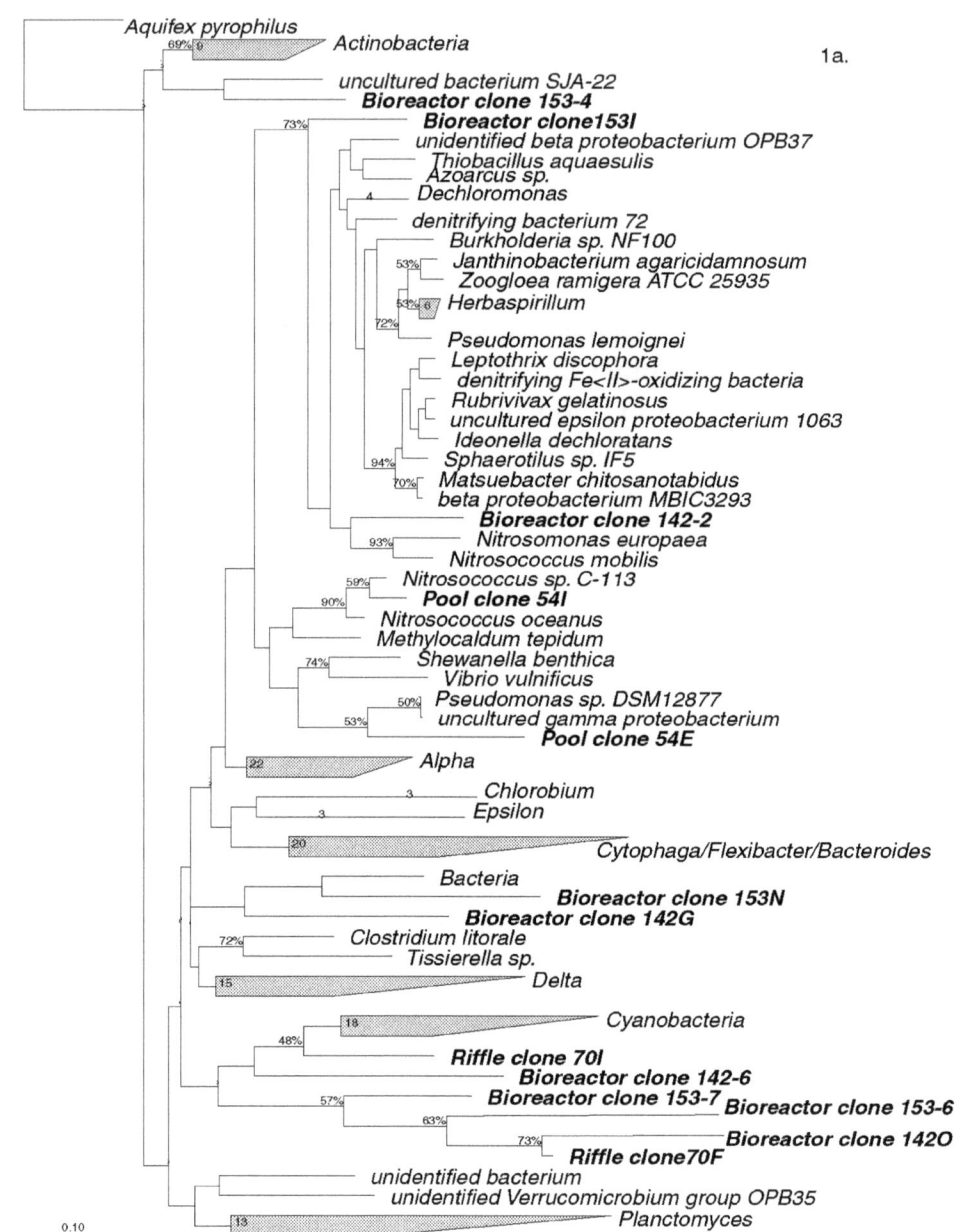

1a.

Figure 3.14 Phylogenetic analysis, using neighbor joining algorithm with bootstrap, of the 16S rDNA genes from environmental samples extracted from epilithon, sediments and bioreactors from the Rio Tempisquito (May 1999). Clones most closely related to the Beta/Gamma Proteobacteria are shown. Boots strap values of greater than 50% are significant.

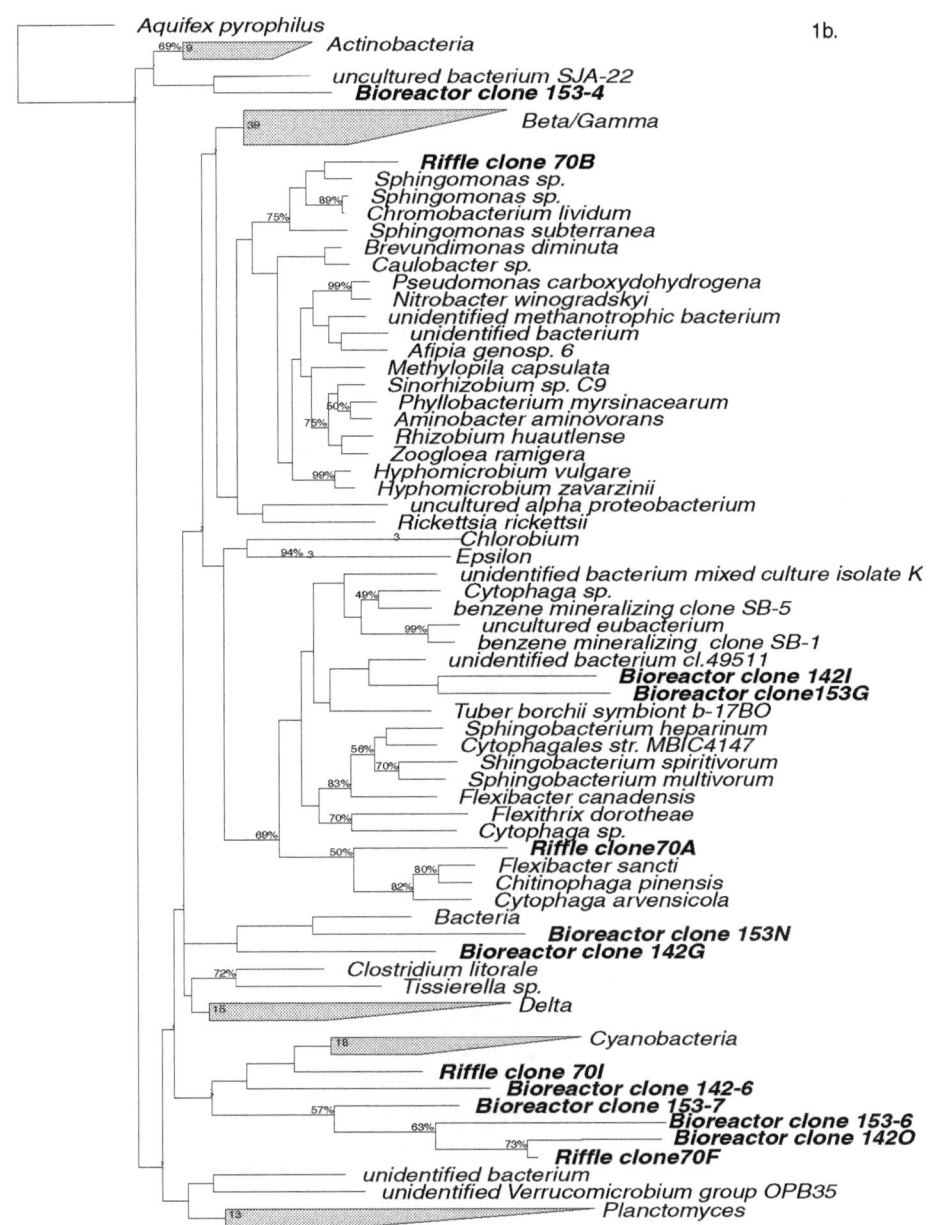

Figure 3.15 Phylogenetic analysis, using neighbor joining algorithm with bootstrap, of the 16S rDNA genes from environmental samples extracted from epilithon, sediments and bioreactors from the Rio Tempisquito (May 1999). Clones most closely related to the Alpha Proteobacteria and Cytophaga/Flexibacteria/Bacterioides are shown. Boots trap values of greater than 50% are significant.

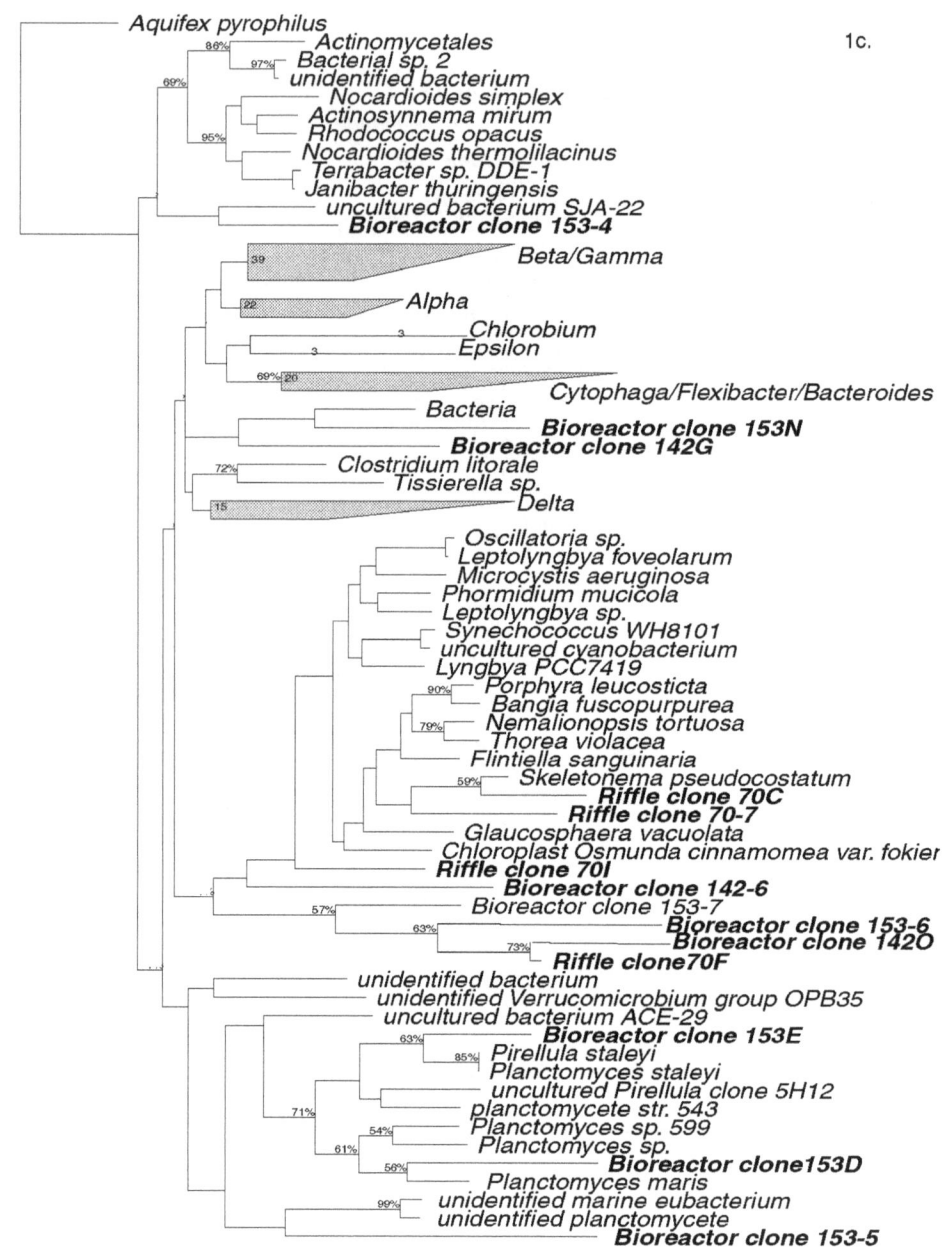

Figure 3.16 Phylogenetic analysis, using neighbor joining algorithm with bootstrap, of the 16S rDNA genes from environmental samples extracted from epilithon, sediments and bioreactors from the Rio Tempisquito (May 1999). Clones most closely related to the Actinobacter, Plantomycetes and Cyanobacteria are shown. Boots strap values of greater than 50% are significant.

64

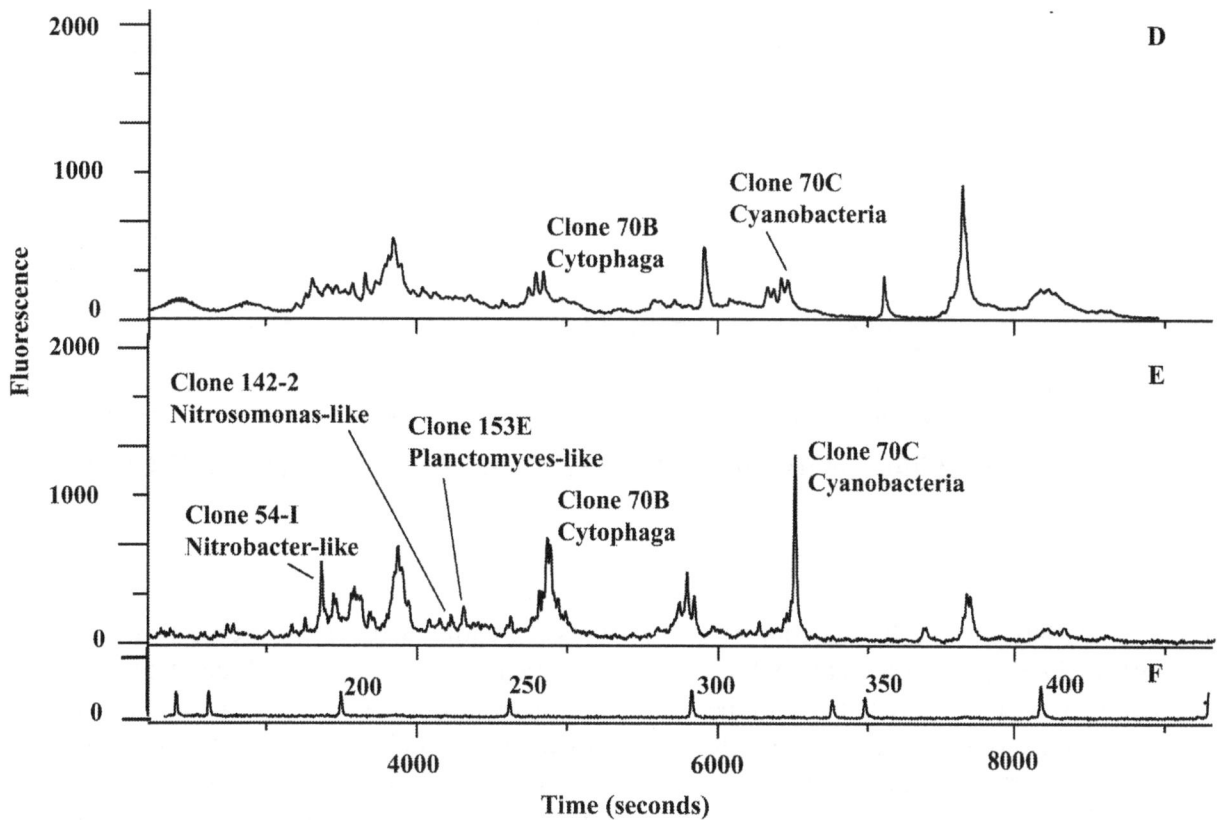

Figure 3.17 The identification of clones in the t-RFLP traces from Rio Tempisquito bioreactor (D) and sediments (E). Panel F is the size standard used to calculate the t-RFLP fragment size.

OBJECTIVE 2B. ABILITY OF BACTERIAL HETEROTROPHS FROM ONE WATERSHED TO METABOLIZE THE BOM PRESENT IN THE WATER FROM ANOTHER WATERSHED

Long-term cross-feeding experiments were used to test the hypothesis that bacterial communities are highly specialized to process polymeric NOM. We hypothesized that the communities that developed in the Betasso WTP and Hillsborough River bioreactors would not have the ecological versatility to metabolize organic matter from White Clay Creek without a significant shift in community structure. Fully colonized bioreactors from the two different watersheds were sent to the Stroud Water Research Center so they could be fed Balston-filtered stream water and sterilized stream water from White Clay Creek. Both sets of biorectors were physically disturbed during shipping as they had a head space of water above the Siran packing, rather than having the chromatography columns fully packed with Siran. The Siran was allowed to settle in the bioreactors prior to exposure to White Clay Creek water. The Betasso WTP bioreactors were exposed to White Clay Creek water for 7 months, and the sterilization of the water was achieved by autoclaving. The Hillsborough River bioreactors were exposed to White Clay Creek water for one month, and the sterilization was achieved by filtration.

65

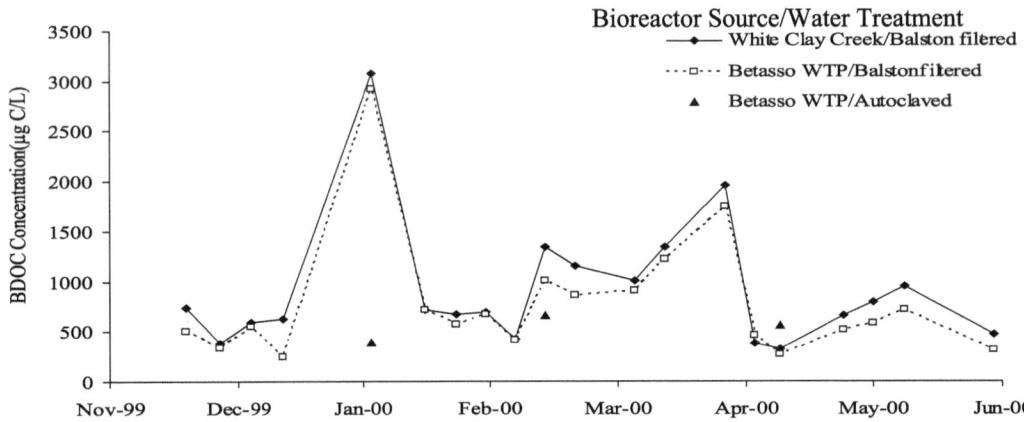

Figure 3.18 Ability of bioreactors from Betasso WTP and WCC to metabolize WCC stream-water DOC

The Barker Lake bioreactors from the Betasso treatment plant were able to metabolize some of the DOC present in White Clay Creek, but typically they were less efficient compared to the White Clay Creek bioreactor, with the sterile treatment slowly becoming more adept at metabolism after mid-March (Figure 3.18). Because autoclaving is such a harsh treatment, these data need to be considered with caution. However, on the last day of exposure, all bioreactors received the same Balston filtered water, and under that circumstance, the differences between the bioreactors from the two sites were clear.

The Hillsborough River bioreactors, with their high-carbon, black-water source, were initially unable to metabolize the White Clay Creek DOC (Table 3.15). They did develop some metabolic ability for that carbon source after one month of exposure, but both the sterile and the non-sterile treatments never approached the levels of the White Clay Creek bioreactors.

Analysis of the t-RFLP samples collected before and after exposure to White Clay Creek water revealed that the bioreactor communities from Barker Lake and the Hillsborough River converged to similar t-RFLP patterns after exposure to White Clay Creek water (75% similarity; Figure 3.19). Prior to exposure to White Clay Creek water, the t-RFLP patterns in the Barker Lake and Hillsborough River bioreactors did not cluster together (0%).

Natural organic matter (NOM) was shown to significantly affect the genetic composition of the microbial communities. Upon exposure to White Clay Creek water, t-RFLP patterns from two very different water treatment systems partially converged. These experiments involved long-term exposure of the Hillsborough River (FLA) and Barker Lake (COL) bioreactors to White Clay Creek water. The data suggest that organic matter composition exerted control over the overall composition of the microbial community, although the White Clay Creek water was not adjusted to mimic inorganic chemical matrix of the Hillborough County or Barker Lake feed-water (i.e. pH, inorganic nutrients, and ionic strength). The native microorganisms that degraded Hillsborough River (FLA) or Barker Lake (COL) NOM could not be competitively maintained in the bioreactor upon exposure to the NOM from the White Clay Creek watershed as shown by a reduction in peak 237 (Figure 3.20). It is possible that the physical disturbance of the bioreactors during shipment gave the White Clay Creek bacteria a chance to invade and colonize the Hillsborough River and Barker Lake bioreactor communities. Other studies have shown that NOM can select for different microbial groups. In a study using fractionated NOM, different

Table 3.15
Ability of bacteria from the Hillsborough River to metabolize BOM from White Clay Creek (all data as μgC/L)

	Water treatment								
	Balston filtered White Clay Creek water						Sterile Filtered White Clay Creek water		
	White Clay Creek bioreactor			Hillsborough River bioreactor			Hillsborough River bioreactor		
	Inflow	Outflow	BDOC [%]	Inflow	Outflow	BDOC [%]	Inflow	Outflow	BDOC [%]
16NOV00	2,554	1,540	1,014	2,554	3,021	0	2,393	2,225	168
			[39.7]			[0]			[7.0]
22NOV00	1,618	845	733	1,618	1,220	398	1,315	1,179	136
			[47.8]			[24.6]			[10.4]
30NOV00	1,533	1,061	472	1,533	1,387	146	1,456	1,267	188
			[30.8]			[9.5]			[12.9]
07DEC00	1,148	689	459	1,148	945	204	1,002	856	146
			[40.0]			[17.7]			[14.6]
14DEC00	1,358	765	593	1,120	865	255	1,050	981	169
			[43.6]			[22.8]			[16.1]

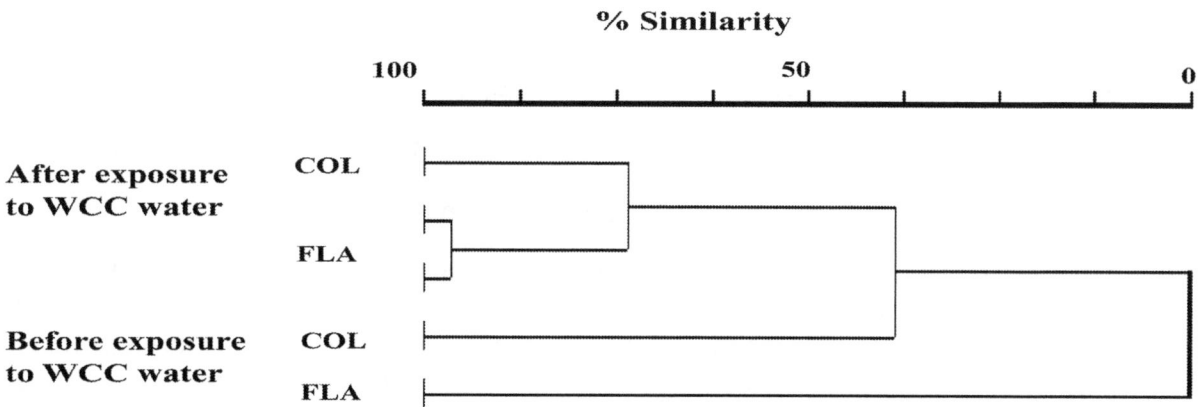

Figure 3.19 The response of the microbial community, measured using t-RFLP analysis, from one watershed to short term exposure to organic matter from a different watershed. There was convergence of t-RFLP patterns after communities that developed in bioreactors from Hillborough River, FL (FLA) and Barker Lake, CO (COL) were exposed to water from White Clay Creek, PA (WCC).

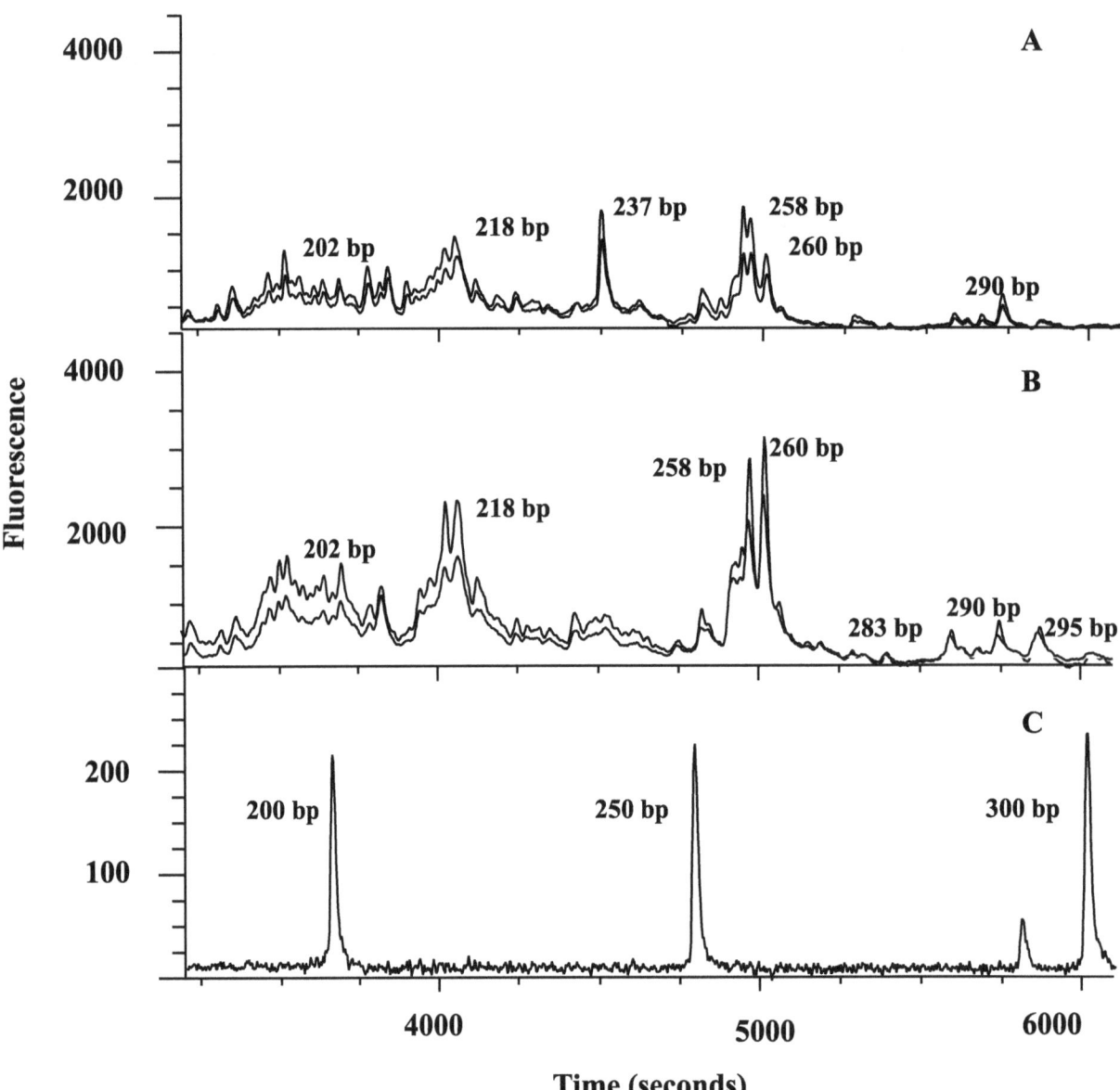

Figure 3.20 A. t-RFLP patterns from two replicate bioreactors colonized by bacteria from the Hillsborough River, FL, and fed water from the Hillsborough River, FL; B. t-RFLP patterns from two replicate bioreactors colonized by bacteria from the Hillsborough River, FL, and fed water from White Clay Creek, PA; C. internal standard used in t-RFLP analysis.

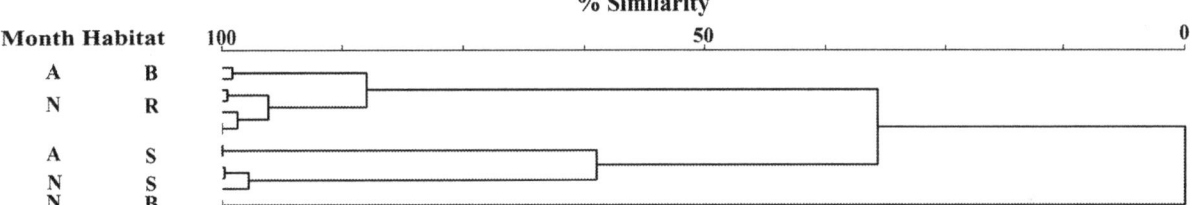

Figure 3.21 The bacterial community response to seasonal temperature and DOM changes in the sediment (S), bioreactor, (B), and epilithon (R) communities in White Clay Creek. Months are denoted by April (A) and November (N).

phylogenetic groups were selectively enriched after long-term exposure to different components of the DOM pool (Covert and Moran 2001). While there were some t-RFLP peaks with a cosmopolitan distribution, high molecular weight NOM was used predominantly by bacteria from the subclasses alpha, delta, and gamma while low molecular weight NOM was used by subclasses gamma and epsilon of the class Proteobacteria. Similarly, in response to an algal bloom, specific bacterial groups were selectively enriched and dominated the community, presumably in response to shifts in NOM availability (Gonzalez et al. 2000; Moeseneder, Winter, and Herndl 2001). While our work has focused on changes in species composition within bacterial communities, future work could address the issue of changes in functional genes.

OBJECTIVE 3A. BACTERIAL COMMUNITY RESPONSE TO SEASONAL CHANGES IN TEMPERATURE AND DOM

Temperature and DOM appeared to have a dramatic affect on the bacterial communities in White Clay Creek. There was a large shift in the communities between April and November (Figure 3.21). However, some peaks were present throughout the study period and others changed seasonally. For example, the relative abundance of peak 105 in White Clay Creek (Figure 3.22) was lower in April than in November. Conversely, several peaks were only present in April or November. Not surprisingly, the response of the bioreactor communities was complex. Although some components of the community responded via increased relative abundance, the majority of the response was the appearance of new t-RFLP peaks in April. The combined effect of seasonal changes in temperature and DOM was a shift in the number of t-RFLP peaks (richness).

OBJECTIVE 3B. BACTERIAL COMMUNITY RESPONSE TO SEASONAL CHANGES IN DOM

When temperature was removed as a variable, shifts in DOM concentration alone had a minimal affect on microbial community composition, as demonstrated by the Rio Tempisquito bioreactors. The bioreactor microbial community was 65% similar during the wet and dry seasons (Figure 3.23). Even though the bioreactors were exposed to a greater number of DOM concentration excursions associated with the rainy season, compared to the dry season (see Appendix A), we observed only a 35% shift (65% similarity) in community composition as shown by t-RFLP analysis. There were more peaks found in common between May and November in Rio Tempisquito

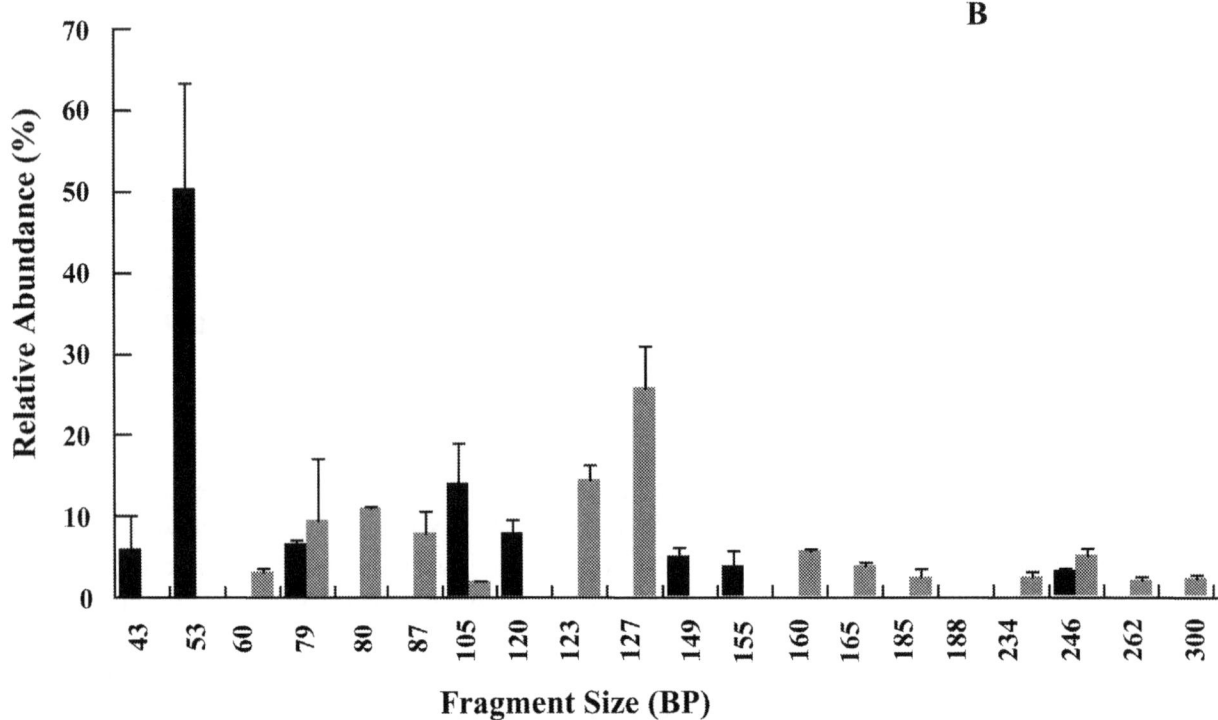

Figure 3.22 Mean and standard deviation (n = 2) of the relative abundance of t-RFLP peaks in White Clay Creek. Black bars are samples taken in November and gray bars are samples taken in April.

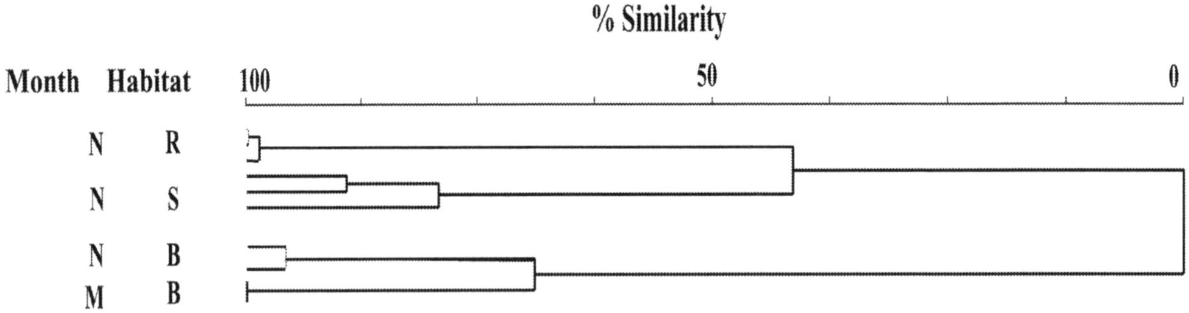

Figure 3.23 The bacterial community response to seasonally constant temperatures and DOM changes in the sediment (S), bioreactor, (B), and epilithon (R) communities in Rio Tempisquito. Months are denoted by May (M) and November (N).

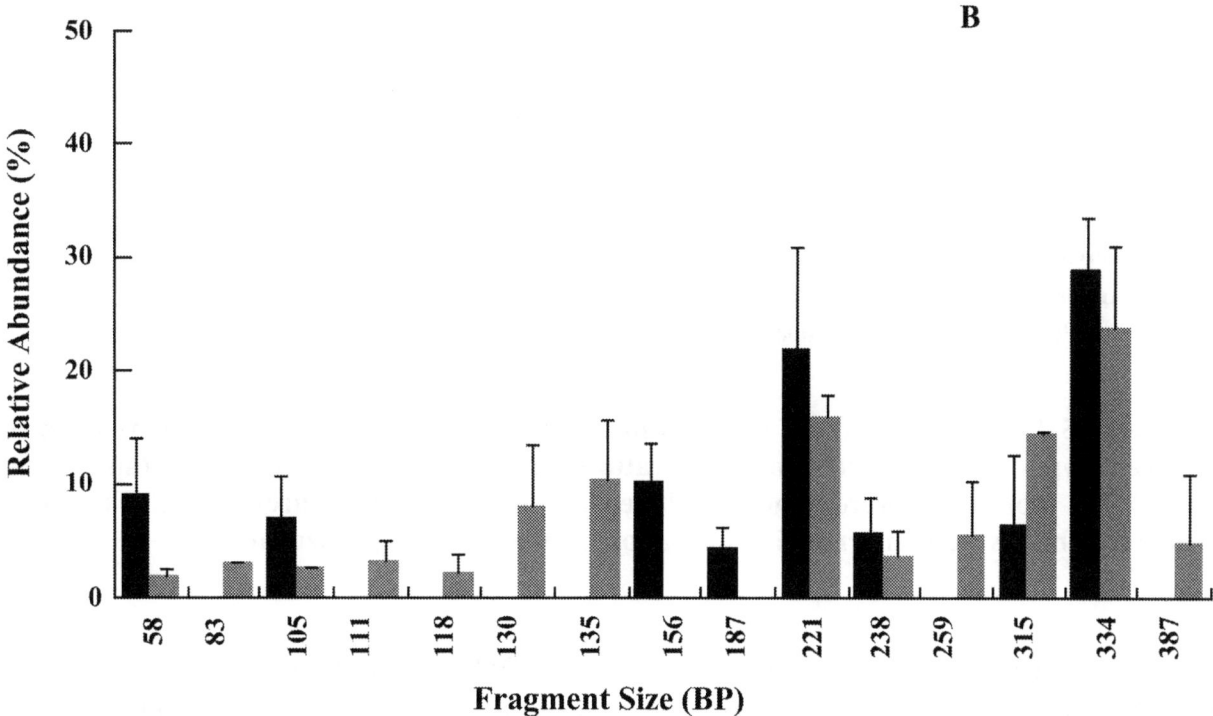

Figure 3.24 Mean and standard deviation (n = 2) of the relative abundance of t-RFLP peaks in Rio Tempisquito. Black bars are samples taken in November and gray bars are samples taken in May.

(Figure 3.24) than observed for the White Clay Creek communities. This suggests that the major response to increased DOM concentration was a selection from within the existing populations rather that a shift in community composition.

OBJECTIVE 3C. BACTERIAL COMMUNITY RESPONSE TO SEASONAL CHANGES IN TEMPERATURE

The combination of molecular analysis of microbial communities with the experimental manipulation of bioreactor systems has identified operating parameters that are likely to be important in the process control of drinking water treatment systems. Temperature and organic matter quality and quantity affected the composition of the microbial community associated with bioreactors and may influence parameters important in the biological treatment of drinking water. The response of the microbial community to these factors was through both recruitment of new species (measured as t-RFLP peaks) and changes in the relative contribution of t-RFLP peaks. The general response to increased temperature appeared to be the appearance of novel t-RFLP peaks. Exposure to novel sources of NOM appeared to also increase novel t-RFLP peaks, whereas shifts in NOM concentration generally appeared to increase the relative abundance of existing t-RFLP peaks. Whether these shifts in population structure were associated with novel metabolic capabilities of the new recruits or shifting metabolic capabilities of existing population is reserved for future research.

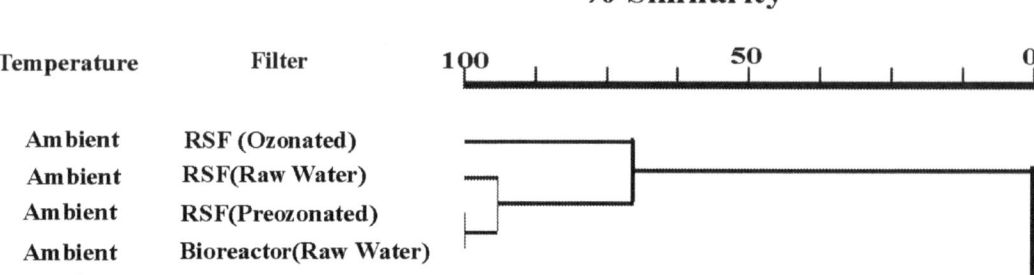

Figure 3.25 Changes in t-RFLP patterns in the microbial communities that developed in different types of filters, different temperatures, and water treatment. RSF (Rapid Sand Filter) and bioreactors were either incubated at ambient temperatures or 3°C. The three water treatments were intermediate ozonation, raw water, and preozonation.

To gauge the shifts in structure and function (DOM metabolism) of the communities that developed in bioreactors and rapid sand filters to temperature, we measured NOM and t-RFLP. In this study, there was very little similarity between the communities that developed on the RSF at 3°C relative to those maintained at ambient temperatures. Additionally, bioreactors at the Betasso water treatment plant were exposed to seasonal variations in NOM and held at either ambient temperature or a constant 9°C. Although there was presumably a seasonal shift in the composition of the DOM, constant low temperature caused a larger shift in the microbial community (Figure 3.25). The influence of temperature for DOM removal at Betasso WTP was evaluated by lowering the temperature of the sand filter through which pre-ozonated water was passed. This DOM (sample 187) was compared to pre-ozonated DOM passed through a room temperature sand filter (sample 190). The effluents were collected and analyzed by TMAH GC MS (Figure 3.26). The differences between two samples were investigated qualitatively and semi-quantitatively by comparing peak and area information on GC MS chromatogram. Overall the distribution of products for the two samples was very similar. In contrast to the response observed from t-RFLP analyses, no notable effect of temperature was found from our chemical analyses of effluents.

The effect of temperature on community function has been associated with changes in community composition. In our study, temperature had an impact on the composition t-RFLP peaks in bioreactors and rapid sand filters (Figure 3.27), although DOC use was unaltered. In contrast, others have noted shifts in t-RFLP patterns across a similar temperature range in soils (Fey and Conrad 2000). The changes in t-RFLP patterns were associated with an alteration of community metabolism. Similar correlations have been noted in wastewater treatment reactors (Kurisu et al. 2002). This suggests that over the range of temperatures tested, different groups of bacteria fill similar functions associated with DOM metabolism within the communities that developed on rapid sand filters.

Influence of temperature for DOM removal at Betasso WTP was evaluated by lowering the temperature of the sand filter through which pre-ozonated water was passed. This DOM (sample 187) was compared to pre-ozonated DOM passed through a room-temperature sand filter (sample 190). The effluents were collected and analyzed by TMAH GC MS (Figure 3.28).

Pre-ozonated water passed through a rapid sand filter at 9°C

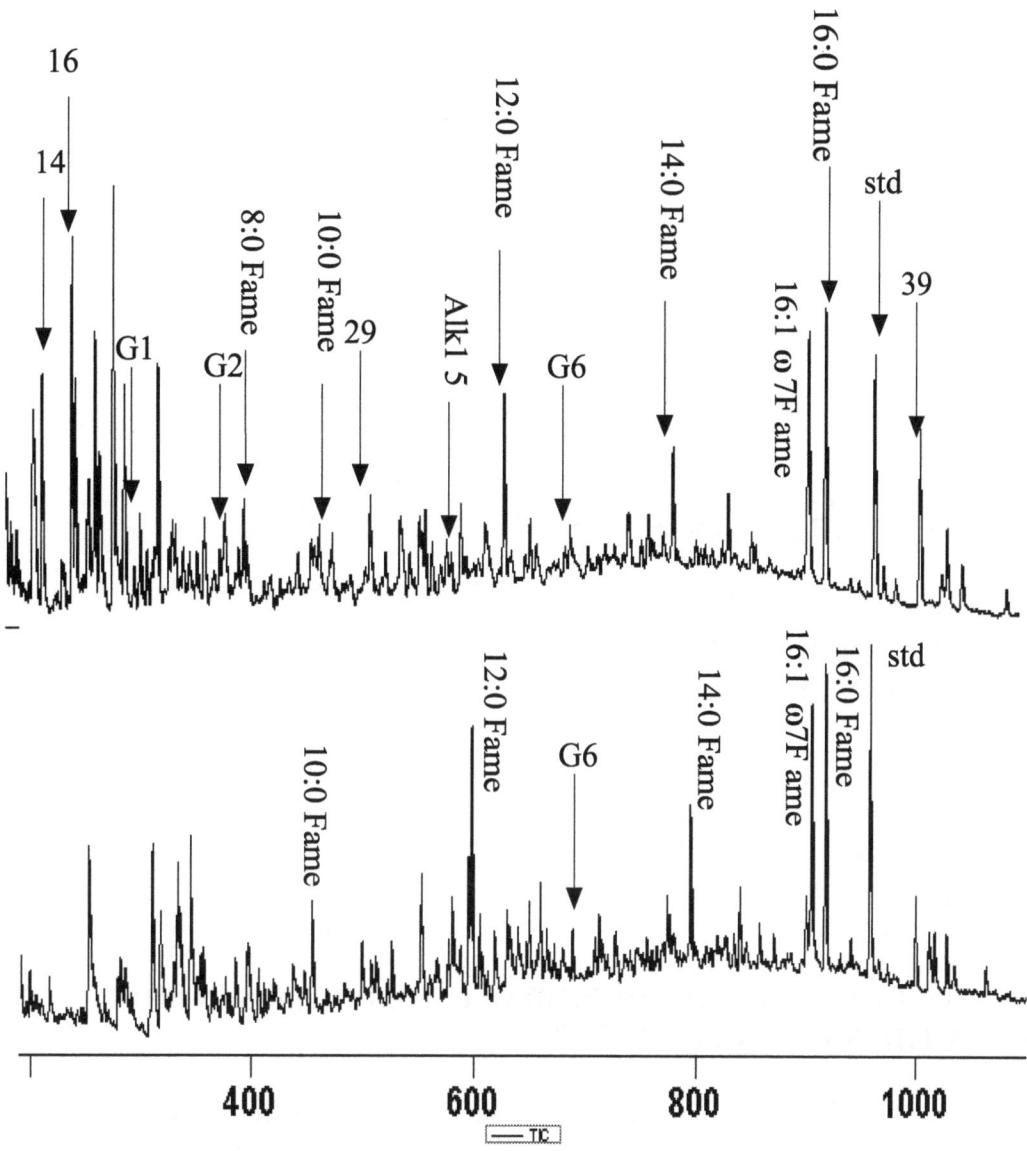

Pre-ozonated water passed through a rapid sand filter at ambient temperature

Figure 3.26 Chromatograms of TMAH thermochemolysis products from Betasso WTP effluent waters from RSF operated at reduced 187 (upper) and ambient temperatures 190 (lower)

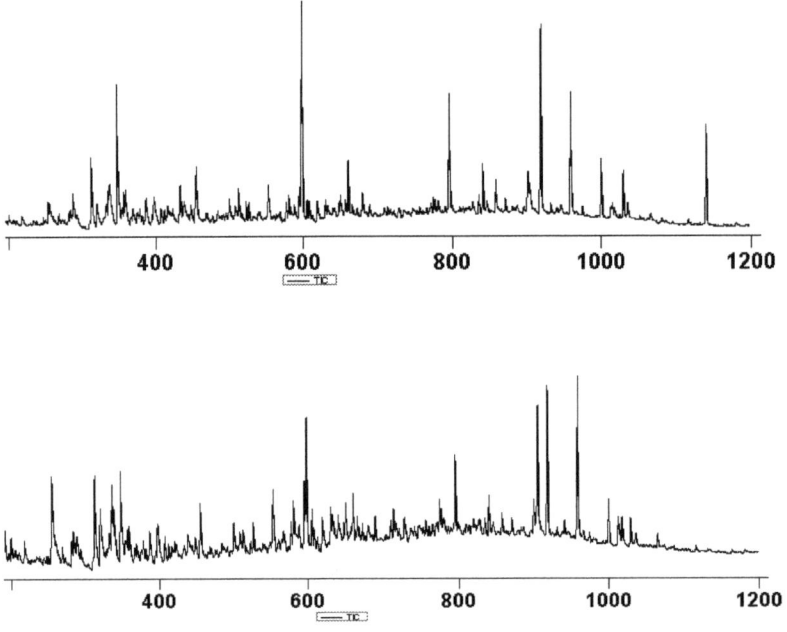

Figure 3.27 Chromatograms of TMAH products in effluents from a bioreactor (upper) and RSF (lower) fed Betasso WTP raw water

Differences between two samples were investigated qualitatively and semi-quantitatively by comparing peak and area information on GC MS chromatogram. Overall the distribution of products for the two samples was very similar. In contrast to the response observed from t-RFLP analyses, no notable effect of temperature was found from our chemical analyses of effluents.

OBJECTIVE 4A. COMPARE THE MICROBIAL COMMUNITIES IN BIOREACTORS TO THOSE IN THE NATURAL HABITATS

The degree to which the bioreactor communities reflected the microbial communities found in the natural system varied between biomes. The Rio Tempisquito bioreactor communities were different from the epilithon and sediment communities (Figure 3.23) regardless of season. While the temperature between bioreactors and the natural environment was similar, other factors, such as flow regime and light that are altered in the bioreactors could have selected for different community members in the bioreactors. Temperature also varied in White Clay Creek between November and April. Seasonal clustering of the sediment communities occurred ranging from 35-62% similarity between seasons (Figure 3.21). In November, the natural community and bioreactors were not similar whereas in April, the similarity between the bioreactors and the sediment community was about 35%. Comparison between the temperate (Figure 3.21) and tropical system (Figure 3.23) suggested that although temperature strongly influenced how well the bioreactors reflected the microbial community in the source waters other factors were also important in tropical systems. Cell densities in the sediment habitats of White Clay Creek (4.9×10^9 cells/gdw) and Rio Tempisquito (3.9×10^9 cells/gdw) were similar to those in the bioreactors. Epilithon densities, expressed on a unit area basis, cannot be directly compared to the bioreactor densities.

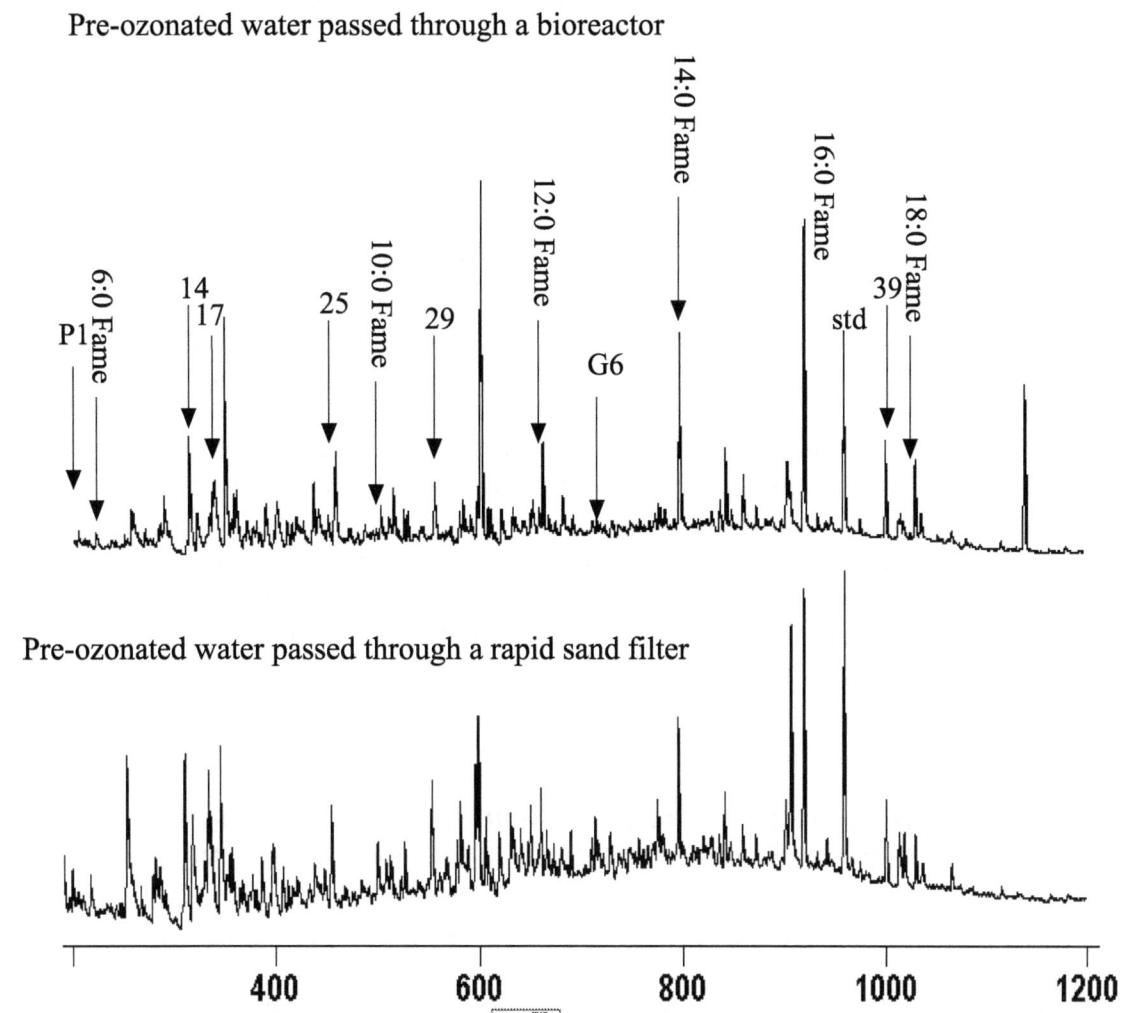

Figure 3.28 Chromatograms for TMAH thermochemolysis products from effluents of a bioreactor (upper) and RSF (lower) fed Betasso WTP pre-ozonated water

OBJECTIVE 4B. COMPARE THE MICROBIAL COMMUNITIES IN BIOREACTORS TO THOSE IN TREATMENT FILTERS

To identify elements that influenced the refinement of the drinking water treatment processes, we used DNA based measures of microbial population structure as a tool to relate changes in microbial community composition to different operating conditions and associated process data. Varying conditions of operation included seasonal shifts in temperature and NOM quantity and quality, as controlled by changing source water or ozonation of source water. The response of the microbial community to these treatments was seen both as changes in the abundance of existing t-RFLP peaks and appearance on new t-RFLP peaks.

Pretreatment of drinking water may have influenced the quality of the organic matter that was available for microbial consumption and thus impacted community composition and the quality of the processed water. In a comparison of two different bioreactors from Betasso WTP, a rapid sand filter and a bioreactor filled with sintered borosilicate beads, differences in community structure were associated with temperature regime and water pretreatment (Figure 3.25). A reduction in temperature, to 3°C, resulted in colonization of an entirely different microbial community in rapid sand filters (RSF 3°C). In contrast, bioreactors and rapid sand filters, which were incubated at ambient temperature, had about 85% similarity in t-RFLP peaks. Ozonation resulted in a shift in community composition by about 30% (70% similarity) when compared to the bioreactor and rapid sand filter incubated at ambient temperatures. Bacterial biomass, determined from phospholipid-P concentrations, expressed as nmoles P/gdw, were higher in the RSF (range 26 to 44) than in the bioreactors (range 11 to 17).

Pretreatment of drinking water is expected to alter the quality of the organic matter available for microbial consumption and thus have an impact on community composition. Both temperature and pretreatment (by ozonation) affected the communities that developed on the rapid sand filters and the bioreactors. Temperature was associated with the most extreme shift in community structure (3°C, RSF, Figure 3.25). In contrast, when incubated at ambient temperature and receiving the same source water, the RSF and bioreactors had about 85% of the t-RFLP peaks. Thus, temperature was shown to be one of the important parameters related to community composition and therefore is implicated as important in controlling the performance of biological filters. The quality of the DOM is another factor that may control the performance of bioreactors as models for biological treatment of drinking water. Quality of DOM was altered using ozonation. Intermediate ozonation resulted in a 15% shift (70% similarity) in t-RFLP patterns when compared to rapid sand filters and bioreactors incubated at ambient temperatures.

OBJECTIVE 4C. ASSESS THE ABILITY OF BIOREACTORS TO PREDICT PERFORMANCE OF BIOLOGICAL FILTERS USED FOR WATER TREATMENT

Analyses of the carbohydrate content of the samples collected at University of Colorado as part of a comparison of bioreactors and rapid sand filters were performed. Dissolved carbohydrates accounted for 2.7% of the DOC in the raw water, 36% of the BDOC removed by the raw water bioreactor and 19% of the BDOC removed by the raw water rapid sand filter (Table 3.16). Pre-ozonation made the bioreactor and rapid sand filter performances closer, presumably by making the BDOC more labile. When fed raw water, the bioreactor removed 52% of BDOC that the rapid sand filter removed, and when fed pre-ozonated water, the bioreactor removed 87% of the BDOC removed by the rapid sand filter. Intermediate ozonation improved DOC removal, but not as much pre-ozonation (influent water to the intermediate ozonation rapid sand filter was the effluent from the rapid sand filter fed raw water), and had no impact on further carbohydrate removal compared to the raw water.

Comparison Between Rapid Sand Filter and Bioreactor

The effluent samples of DOM collected from the bioreactor and the sand filter were compared by TMAH GC/MS to assess whether the bioreactor can mimic a rapid sand filter. Two sets of samples were compared. The first set was composed of raw water effluent fed through a bioreactor (sample 188) and through a rapid sand filter (sample 191) and begged the question as to whether a bioreactor can mimic a sand filter for raw input water. The second set was effluent water samples from the pre-ozonation system passed through a bioreactor (sample 189) and through a rapid sand filter at ambient temperature (sample 190). This set addressed the question as to whether pre-ozonation affected the comparison bewteen bioreactor and sand filter.

76

Table 3.16
Comparison of bioreactor and rapid sand filter performance

Sample	Concentration (µgC/L)		
	DOC	BDOC	DTCHO
Feed water			
Raw	2094		56
Pre-ozonated ambient	1808		46
Rapid sand filter effluents			
Intermediate ozonation	1640	260	25
Raw	1900	194	24
Pre-ozonated 3°C	1656	152	34
Pre-ozonated ambient	1318	490	30
Bioreactor effluents			
Raw	1994	100	20
Pre-ozonated ambient	1384	424	24

Quantitative analyses were performed for each set of samples and the results are shown in Tables 3.17 and 3.18. In the case of the first question regarding raw water (Figure 3.27), the GC-MS chromatograms obtained from each of the samples were quite different. This is confirmed by quantitative analysis. The effluent from the rapid sand filter has higher concentrations of lignin-derived compounds. However the concentration of fatty acid derivatives especially dodecanoic acid methyl ester and methyl tetradecanoate are considerably higher for the bioreactor effluent, up to 1000 ng L^{-1} higher for each compound. It is clear that the biorector is adding microbial inputs to the water stream in far excess over what would be present in sand filters, perhaps indicating greater efficacy in bacterial reworking of the DOC.

In case of pre-ozonated water being fed through the bioreactor and through sand filter (Figure 3.28), the difference between two chromatograms was much smaller for both systems, reflected by concentration differences that are smaller (up to 50 ng L^{-1}). It should be pointed out that the overall concentrations of measured compounds were lower than those from equivalent processing of raw water samples. This strongly suggests that both the rapid sand filter and bioreactor may be more efficient in removing NOM when it is combined with pre-ozonation treatment. As a conclusion, raw water passing through a bioreactor is altered in a significantly different manner compared to raw water passing through a rapid sand filter. However, in the case of pre-ozonated water passing through equivalent systems, effluent from the bioreactor showed similar concentration ranges of compounds to those from rapid sand filter.

It is clear that the TMAH thermochemolysis technique provides important molecular-level fingerprints of the DOM passing through various points in water treatments facilities and could be used to assess the efficacy and performance of the systems. The studies commenced as part of this investigation are preliminary in that much of the effort was invested to render the TMAH analysis quantitative and useful, and less effort was invested in a detailed implementation of the technique.

Table 3.17
TMAH quantitative analysis on effluents from bioreactor and rapid sand filter fed by untreated raw water

Compound	Concentrations (ng L^{-1})		
	Bioreactor	Rapid sand filter	Difference
(P1) benzene, methoxy-	21.3	49.5	-28.2
(P2) benzene, 1-methoxy-4-methyl-	23.9	53.8	-29.8
(G1) benzene, 1,2-dimethoxy-	40.3	107.6	-67.2
(G2) 3,4-dimethoxytoluene	19.3	50.8	-31.6
(S1) 1,2,3-trimethoxybenzene	31.7	74.2	-42.5
(P6) benzoic acid, 4-methoxy-, methyl ester	73.0	171.6	-98.6
(S2) benzene, 1,2,3-trimethoxy-5-methyl-	20.5	47.3	-26.8
(G4) benzaldehyde, 3,4-dimethoxy-	17.3	31.5	-14.1
(G5) ethanone, 1-(3,4-dimethoxyphenyl)	52.2	124.8	-72.5
(G6) benzoic acid, 3,4-dimethoxy-, methyl ester	70.0	161.9	-91.9
(S6) benzoic acid, 3,4,5-trimethoxy-, methyl ester	32.3	79.9	-47.5
octanoic acid, methyl ester	124.7	127.2	-2.5
undecanoic acid, methyl ester	176.3	124.8	51.6
dodecanoic acid, methyl ester	1084.0	38.5	1045.6
methyl tetradecanoate	196.8	73.7	123.1
9-hexadecenoic acid, methyl ester, (Z)-	64.1	85.2	-21.2
hexadecanoic acid, methyl ester	96.5	62.1	34.4
9-octadecenoic acid (Z)-, methyl ester	31.5	68.8	-37.3
octadecanoic acid, methyl ester	28.3	19.4	8.9
eicosanoic acid, methyl ester	4.9	9.9	-5.1
heneicosanoic acid, methyl ester	4.5	8.3	-3.9

Table 3.18
TMAH quantitative analysis on effluents from bioreactor and rapid sand filter fed by pre-ozonated water (samples 189 and 190)

Compound	Concentrations (ng L^{-1})		
	Bioreactor	Rapid sand filter	Difference
(P1) benzene, methoxy-	23.4	17.0	6.4
(P2) benzene, 1-methoxy-4-methyl-	26.1	19.6	6.5
(G1) benzene, 1,2-dimethoxy-	43.3	35.3	8.0
(G2) 3,4-dimethoxytoluene	24.3	35.2	-10.9
(S1) 1,2,3-trimethoxybenzene	33.3	28.3	5.0
(P6) benzoic acid, 4-methoxy-, methyl ester	78.8	57.0	21.8
(S2) benzene, 1,2,3-trimethoxy-5-methyl-	28.0	25.1	2.9
(G4) benzaldehyde, 3,4-dimethoxy-	17.0	11.8	5.3
(G5) ethanone, 1-(3,4-dimethoxyphenyl)	58.9	44.5	14.3
(G6) benzoic acid, 3,4-dimethoxy-, methyl ester	72.6	62.1	10.4
(S6) benzoic acid, 3,4,5-trimethoxy-, methyl ester	38.3	27.8	10.5
octanoic acid, methyl ester	62.4	43.3	19.1
undecanoic acid, methyl ester	68.6	45.6	22.9
dodecanoic acid, methyl ester	45.6	32.4	13.2
methyl tetradecanoate	85.9	43.9	42.0
9-hexadecenoic acid, methyl ester, (Z)-	43.3	91.3	-48.1
hexadecanoic acid, methyl ester	94.5	59.3	35.2
9-octadecenoic acid (Z)-, methyl ester	39.5	27.8	11.8
octadecanoic acid, methyl ester	29.4	15.5	13.9
eicosanoic acid, methyl ester	6.2	5.0	1.2
heneicosanoic acid, methyl ester	5.9	5.2	0.8

CHAPTER 4
SUMMARY, CONCLUSIONS AND RECOMMENDATIONS

Investigations into the structure and composition of NOM, BOM, and microbial communities were carried out in 5 different source waters covering a broad range of vegetative zones and NOM concentrations. The application of advanced chemical techniques, including TMAH GC-MS, and microbiological techniques, including t-RFLP, generated an unprecedented data set describing the molecular nature of NOM and BOM and the species composition of heterotrophic bacterial communities. The data included information on raw water and natural communities in the source waters, as well as information on treatment plant filters and effluents.

Structural and compositional studies with TMAH GC-MS identified unique NOM signatures for the different source waters, with some classes of molecules observed only in single sources. The combination of TMAH GC-MS analyses with bioreactor processing revealed that the BOM pool includes humic substances and lignin, sources generally presumed to be resistant to biodegradation. Additional novel insights included the quantitative contribution of aromatic molecules in the BOM pool and the potential for bacterial demethylation of lignin, an activity typically associated only with the activities of white-rot fungi that decompose wood. TMAH thermochemolysis GC-MS was successfully applied to the investigation of the lignin, carbohydrate, and lipid signature of aquatic DOM and when combined with bioreactor measurements, provided a broader description of the BOM pool within drinking waters. Extensive modifications to the method were required to analyze these natural samples, and those modifications will make the procedure useful for other investigators. Furthermore, the quantitative ability of this procedure adds a new dimension to the comparison of DOM samples within and between laboratories.

Our application of molecular microbiological techniques required optimization of existing molecular protocols to minimize inhibition of enzymes by compounds that naturally co-occur with DNA extracted from different matrix materials. An important part of this research was the optimization of methods at several levels of data gathering and analysis. We also developed and optimized two computer programs that enhanced data analysis. These analytical advances should allow other researchers to successfully analyze other samples of interest and extend the limited data set on microbial community composition in drinking water sources.

The communities of microorganisms that developed within natural streambed habitats and in bioreactors were uniquely different among watersheds. While microbial communities within biorectors fed streamwater differed from the natural communities, especially with regard to the algal and cyanobacterial components in the natural environments, bacterial clones found in the streams were also found in the bioreactors. NOM influenced the genetic composition of microbial communities that developed in bioreactors, and seasonal shifts were observed for watersheds possessing strong seasonal temperature signals. Thus, temperature and organic matter quantity and quality likely influence parameters important to the biological treatment of drinking water. A comparison of bioreactor and rapid sand filters showed some overlap, suggesting the bioreactors may indicate ultimate potential of rapid sand filters for BOM processing.

Using molecular techniques, we showed that we could measure bioreactor communities reproducibly. We also identified bacteria found in both the bioreactor and the sediment communities that were phylogenetically related to organisms that are known to metabolize a wide array of organic carbon. Temperature was shown to be a key factor in bioreactor operation. While keeping the DOM source constant, changes in temperature altered community structure but not DOM consumption. Apparently, different species of bacteria under different temperature regimes fulfill similar functions associated with DOC metabolism. Organic matter also exerted control over the overall composition of the microbial community and metabolism. While keeping temperature regimes constant, shifting the source of DOM fed to a bioreactor community altered both the existing community structure and function (DOM consumption).

CONCLUSIONS AND RECOMMENDATIONS

The data obtained from TMAH and t-RFLP methodologies support our contention that source waters can possess NOM and BOM of widely different quantity and quality, and that these differences influence the community composition of heterotrophic bacteria that use BOM as a source of carbon and energy. The exploratory nature of our research makes it difficult to prescribe specific treatment steps, but we can recommend the following:

- The species composition of bioreactors and their ability to biodegrade DOC specific to a water source should be recognized when considering bioreactor implementation for monitoring BOM concentrations.
- Bioreactors designed to monitor a BOM source ideally should be inoculated, colonized and maintained by that source; at a minimum, acclimation to the source over several months is needed.
- Bioreactors have potential to aid in water treatment as a means of monitoring BOM concentrations.
- Bioreactors installed and operated in a treatment plant filter gallery should be used to follow the changes in BOM during the treatment process to allow for adjustment in the treatment processes.
- Seasonally driven changes in BOM composition and microbial community composition should be considered when designing biologically active filters.
- Seasonal changes in the microbial community colonizing a biologically active filter may diminish filter performance and require an acclimation period to restore perfomance.
- Molecular-based methods for both microbial and chemical analyses of drinking water and treatment processes should be targeted for continued development and implementation within the drinking water industry.

APPENDIX A
TEMPORAL VARIATION IN DOC AND BDOC
AT EACH OF THE STUDY SITES

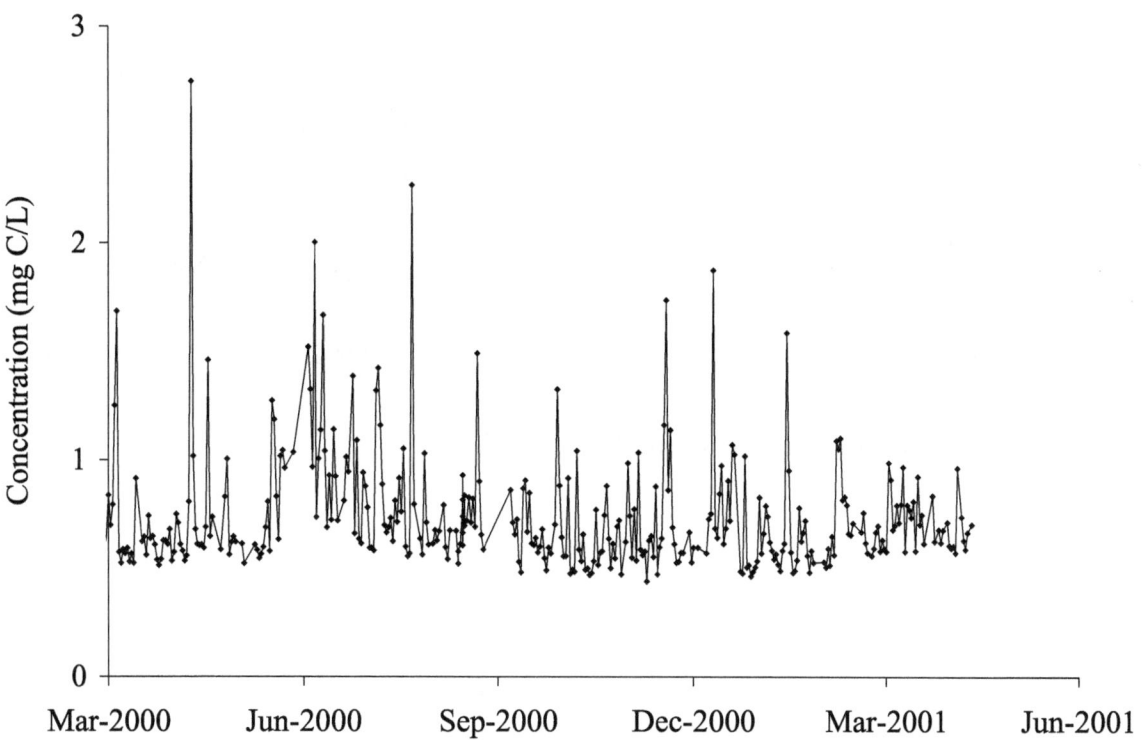

Figure A.1 DOC concentrations in Rio Tempisquito

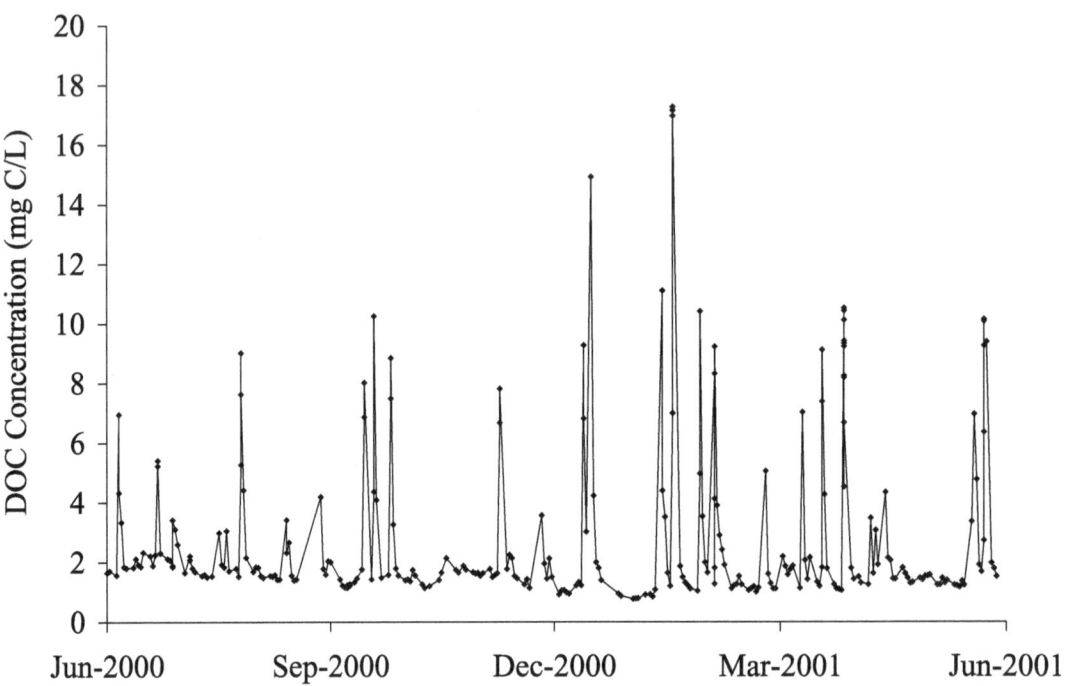

Figure A.2 DOC concentrations in White Clay Creek

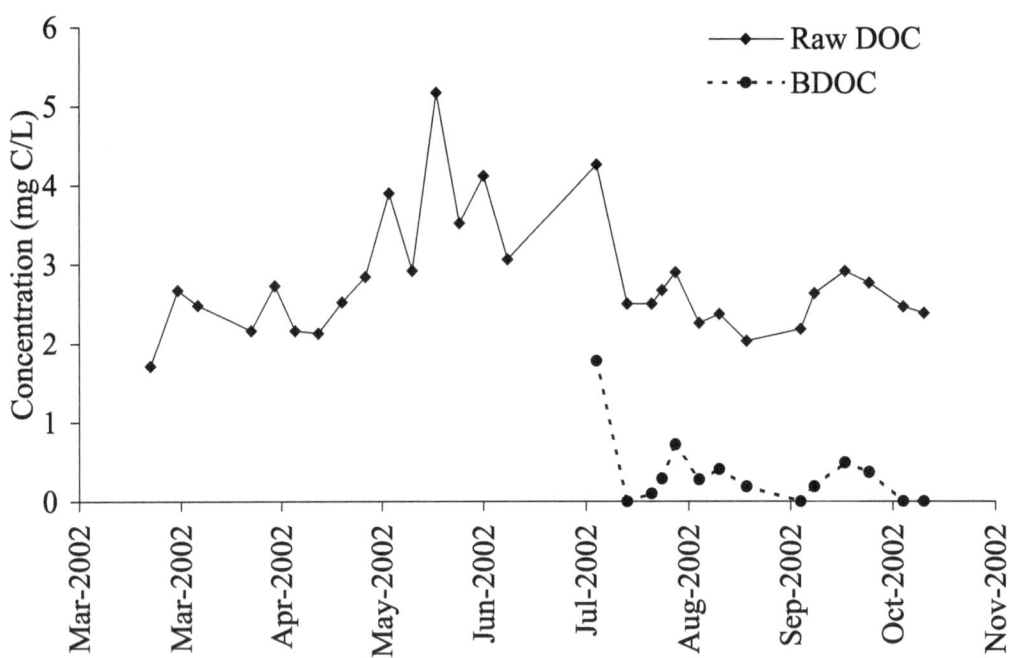

Figure A.3 DOC and BDOC concentrations in Barker Lake

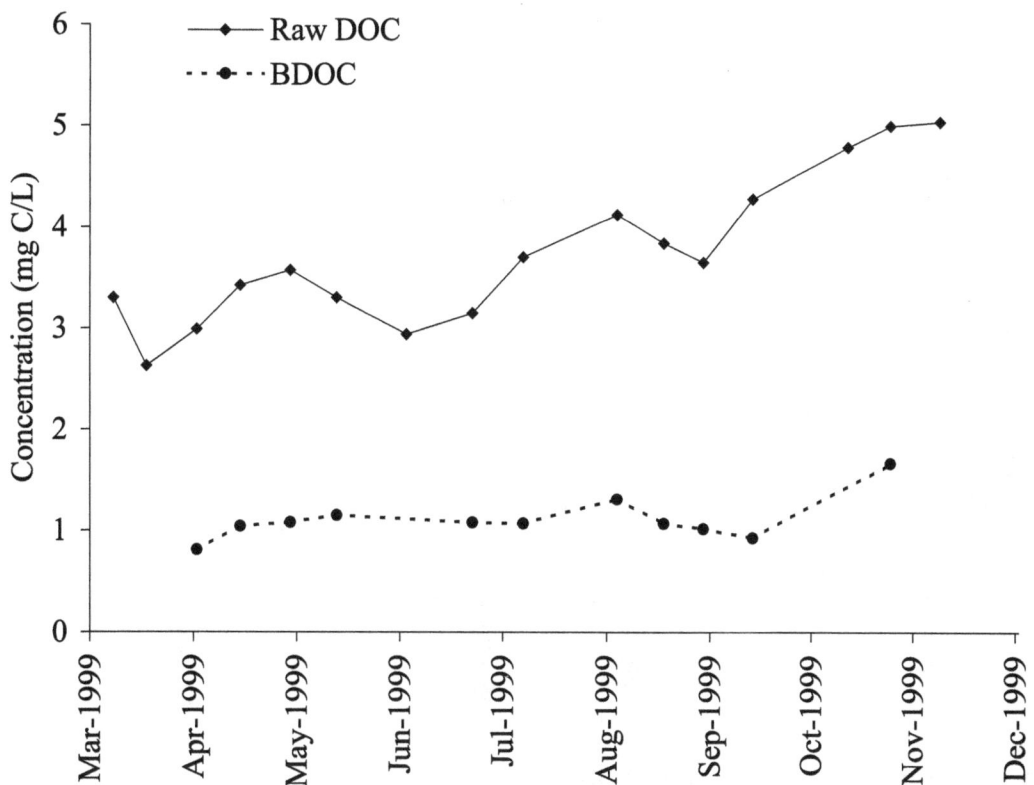

Figure A.4 DOC and BDOC concentrations in White River

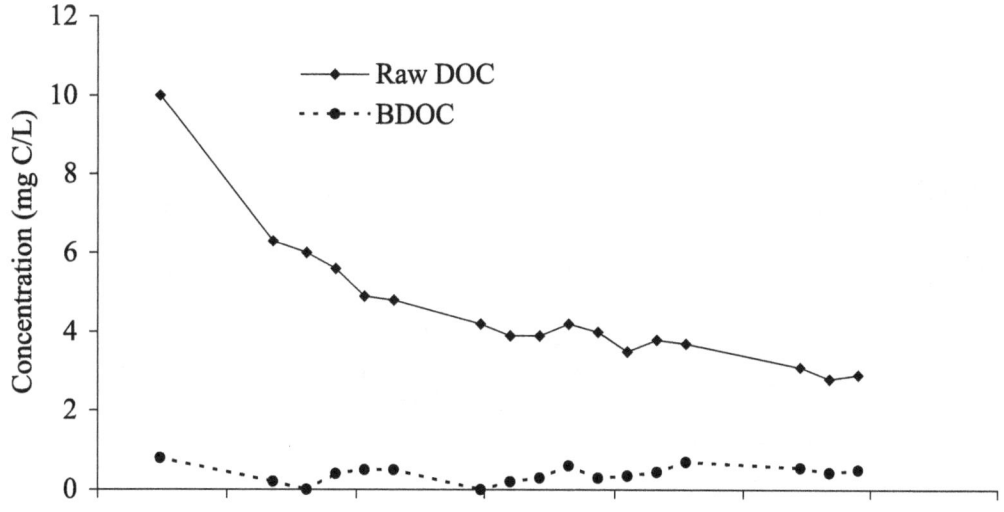

Figure A.5 DOC and BDOC concentrations in Hillsborough River

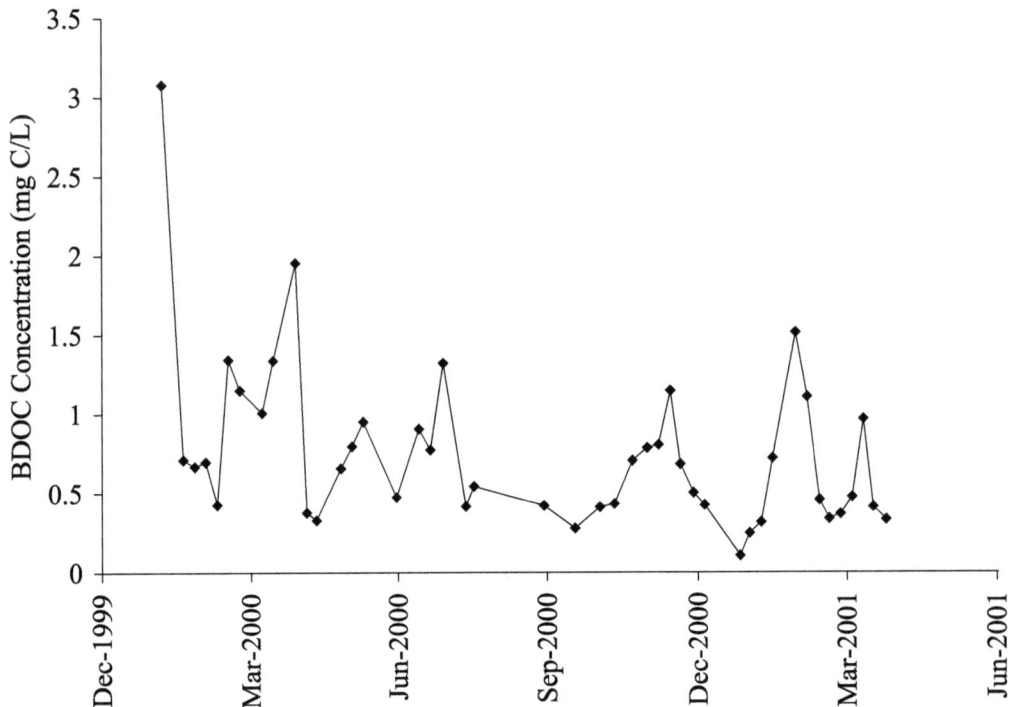

Figure A.6 BDOC concentrations in White Clay Creek

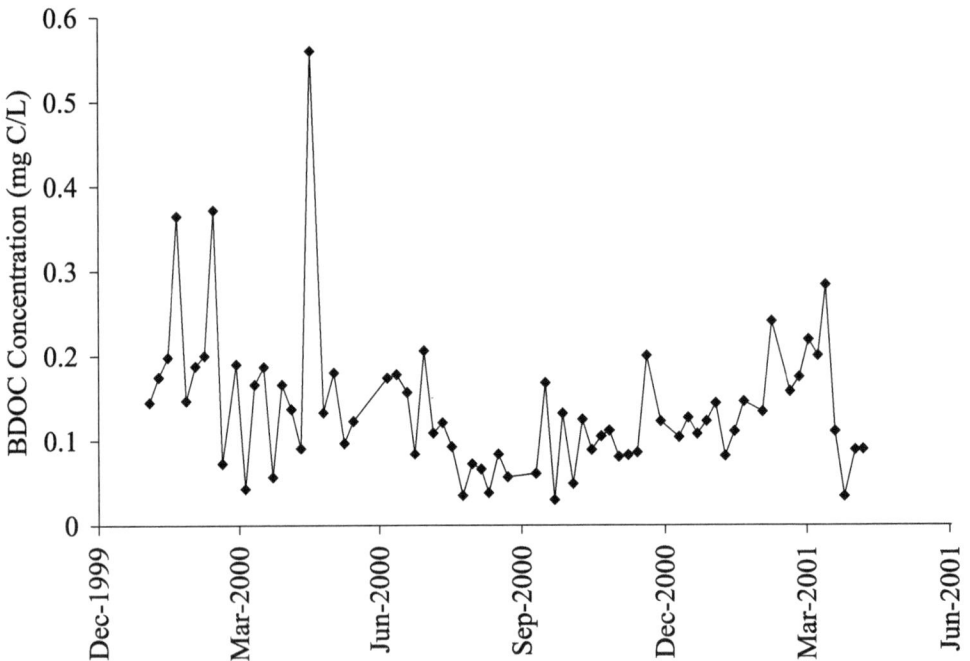

Figure A.7 BDOC concentrations in Rio Tempisquito

REFERENCES

Aiken, G.R., E.M. Thurman, R.L. Malcolm, and H.F. Walton. 1979. Comparison of XAD Macroporous Resins for the Concentration of Fulvic Acid From Aqueous Solution. *Analytical Chemistry*, 51:1799-1803.

Aluwihare, L.I., and D.J. Repeta. 1999. A Comparison of the Chemical Characteristics of Oceanic DOM and Extracellular DOM Produced by Marine Algae. *Marine Ecology Progress Series*, 186:105-117.

Amann, R., W. Ludwig, and K.-H. Schleifer. 1994. Identification of Uncultured Bacteria: A Challenging Task for Molecular Taxonomists. *ASM News*, 60:360-365.

Amann, R.I., W. Ludwig, and K.-H. Schleifer. 1995. Phylogenetic Identification and *in situ* Selection of Individual Microbial Cells Without Cultivation. *Microbiological Reviews*, 59:143-169.

Anderson, W.B., D. Urfer, and P.M. Huck. 1997. Application of a Bench-Scale BDOC Column Method to Full-Scale Plant Conditions. AWWA Water Quality Technology Conference, November 9-12, 1997. Denver, Colo.: AWWA.

Bott, T.L., and L.A. Kaplan. 1993. Persistence of a Surrogate for a Genetically Engineered Cellulolytic Microorganism and Effects on Aquatic Community and Ecosystem Properties: Mesocosm and Stream Comparisons. *Canadian Journal of Microbiology*, 39:686-700.

Butterfield, P.W., B. Ellis, A. Camper, and W. Jones. 1997. Evaluation of Growth Kinetics of Model Drinking Water System Biofilm Cells Utilizing Humic Substances as the Primary Carbon Sources. AWWA Water Quality Technology Conference, November 9-12, 1997. Denver, Colo.: AWWA.

Chefetz, B., J.D.H. van Heemst, Y. Chen, C.P. Romaine, J. Chorover, R. Rosario, G. Mingxin, and P.G. Hatcher. 2000a. Organic Matter Transformations During the Weathering Process of Spent Mushroom Substrate. *Journal of Environmental Quality*, 29:592-602.

Chefetz, B., Y. Chen, C.E. Clapp, and P.G. Hatcher. 2000b. Characterization of Organic Matter in Soils by Thermochemolysis Using Tetramethylammonium Hydroxide (TMAH). *Soil Science Society of America Journal*, 64(2):583-589.

Covert, J.A., and M.A. Moran. 2001. Molecular Characterization of Estuarine Bacterial Communities That Use High and Low Molecular Weight Fractions of Dissolved Organic Carbon. *Aquatic Microbial Ecology*, 25:127-139.

Crawford, R.L. 1981. *Lignin Biodegradation and Transformation*. New York: Wiley.

del Rio, J.C., and P.G. Hatcher. 1996. Structure Characterization of Humic Substance Using Thermochemolysis With Tetramethylammonium Hydroxide. In *Humic and Fulvic Acids. Isolation, Structure, and Environmental Role*. Edited by J.S. Gaffney, N.A. Marley, and S.B. Clark. ACS Symposium Series 651, Washington, D.C.: American Chemical Society. pp. 78-95.

del Rio, J.C., D.E. McKinney, H. Knicker, M.A. Nanny, R.D. Minard, and P.G. Hatcher. 1998. Structural Characterization of Bio- and Geo-Macromolecules by Off-Line Thermo-chemolysis With Tetramethylammonium Hydroxide. *Journal of Chromatography A*, 823:433-448.

Dittman, A., H.T.P. Quinn, and G.A. Nevitt. 1996. Timing of Imprinting to Natural and Artificial Odors by Coho Salmon (*Oncorhynchus kisutch*). *Canadian Journal of Fisheries and Aquatic Sciences*, 53:434-442.

Don, R.H., P.T. Cox, B.J. Wainwright, K. Baker, and J.S. Mattick. 1991. Touchdown PCR to Circumvent Spurious Priming During Gene Amplification. *Nucleic Acids Research*, 19:4008-4010.

Dunbar J., L.O. Ticknor, and C.R. Kuske. 2000. Assessment of Microbial Diversity in Four Southwestern United States Soils by 16S rRNA Gene Terminal Restriction Fragment Analysis. *Applied and Environmental Microbiology*, 66:2943-2950.

Ellis, B.D., P. Butterfield, W.L. Jones, G.A. McFeters, and A.K. Camper. 1999. Effects of Carbon Source, Carbon Concentration, and Chlorination on Growth Related Parameters of Heterotrophic Biofilm Bacteria. *Microbial Ecology*, 38(4):330-347.

Ertel, J.R., and J.I. Hedges, 1984. The Lignin Component of Humic Substances: Distribution Among Soil and Sedimentary Humic, Fulvic, and Base-Insoluble Fractions. *Geochimica Cosmochimica Acta*, 48(10):2065-2074.

Ertel, J.R., J.I. Hedges, and E.M. Perdue. 1984. Lignin Signature of Aquatic Humic Substances. *Science* (Washington, D. C., 1883-), 223(4635):485-487.

Fabbri, D., and R. Helleur. 1999. Characterization of the Tetramethylammonium Hydroxide Thermochemolysis Products of Carbohydrates. *Journal of Analytical and Applied Pyrolysis*, 49:277-293.

Fey, A., and R. Conrad. 2000. Effect of Temperature on Carbon and Electron Flow on the Soil Acrshaeal Community in Methnogenic Rice Field Soil. *Applied and Environmental Microbiology*, 66(11):4790-4797.

Filley, T.R., P.G. Hatcher, W.C. Shortle, and R.T. Praseuth. 2000. The Application of 13C-Labeled Tetramethylammonium Hydroxide (13C-TMAH) Thermochemolysis to the Study of Fungal Degradation of Wood. *Organic Geochemistry*, 31(2-3):181-198.

Filley, T.R., R.D. Minard, and P.G. Hatcher. 1999. Tetramethylammonium Hydroxide (TMAH) Thermochemolysis: Proposed Mechanisms Based Upon the Application of 13C-Labeled TMAH to a Synthetic Model Lignin Dimer. *Organic Geochemistry*, 30(7):607-621.

Findlay, R.H., G.M. King, and L. Watling. 1989. Efficacy of Phospholipid Analysis in Determining Microbial Biomass in Sediments. *Applied and Environmental Microbiology*, 54:2888-2893.

Findlay, S., G.E. Likens, L. Hedin, S.G. Fisher, and W.H. McDowell. 1997. Organic Matter Dynamics in Bear Brook, Hubbard Brook Experimental Forest, New Hampshire, USA. *Journal of the North American Benthological Society*, 16(1):43-46.

Fogel, G.B., C.R. Collins, J.Li, and C.F. Brunk. 1999. Prokaryotic Genome Size an SSU rDNA Copy Number: Estimation of Microbial Relative Abundance From a Mixed Population. *Microbial Ecology*, 38:93-113.

Fonseca, C., R.S. Summers, and M.T. Hernandez. 2001. Comparative Measurements of Microbial Activity in Drinking Water Biofilters. *Water Research*, 35(16):3817-3824.

Frazier, S.W. 2001. The Qualitative and Quantitative Analysis of Organic Matter From Natural Waters Using Tetramethylammonium Hydroxide (TMAH) Thermochemolysis GC-MS. Master's thesis. The Ohio State University.

Frazier, S.W., K.O. Nowack, K.M. Goins, F.S. Cannon, L.A. Kaplan, and P.G. Hatcher. 2003. Characterization of Organic Matter From Natural Waters Using Tetramethylammonium Hydroxide Thermochemolysis GC-MS. *Journal of Analytical and Applied Pyrolysis*, 70:99-128.

Fuhrman, J.A., K. Mccallum, and A.A. Davis. 1993. Phylogenetic Diversity of Subsurface Marine Microbial Communities From the Atlantic and Pacific Oceans. *Applied and Environmental Microbiology*, 59:1294-1302.

Fulthorpe, R.R., A.N. Rhodes, and J.M. Tiedje. 1996. Pristine Soils Mineralize 3-Chlorobenzoate and 2,4-Dichlorophenoxyacetate via Different Microbial Populations. *Applied and Environmental Microbiology*, 62:1159-1166.

Giovannoni, S.J., E.F. DeLong, G.J. Olsen, and N. R. Pace. 1988. Phylogenetic Group Specific Oligodeoxynucleotide Probes for Identification of Single Microbial Cells. *Journal of Bacteriology*, 170:720-726.

Gonzalez, J.M., R. Simo, R. Massana, J.S. Covert, E.O. Casamaor, C. Pedros-alio, and M.A. Moran. 2000. Bacterial Community Structure Associated With a Dimethylsulfoniopropionate-Producing North Atlantic Algal Bloom. *Applied and Environmental Microbiology* 66(10):4237-4246.

Gremm, T.J., and L.A. Kaplan. 1997. Dissolved Carbohydrates in Streamwater Determined by HPLC and Pulsed Amperometric Detection. *Limnology and Oceanography*, 42:385-393.

Hasler, A.D., and W. Wisby. 1951. Discrimination of Stream Odour by Fishes and Its Relation to Parent Stream Behavior. *American Naturalist*, 85:223-238.

Hatcher, P.G. 1987. Chemical Structural Studies of Natural Lignin by Dipolar Dephasing Solid-State 13C Nuclear Magnetic Resonance. *Organic Geochemistry*, 11:31-39.

Hatcher, P.G., and D.J. Clifford. 1994. Flash Pyrolysis and *in situ* Methylation of Humic Acids From Soil. *Organic Geochemistry*, 21:1081-1092.

Hatcher, P.G., and R.D. Minard. 1996. Comparison of Dehydrogenase Polymer (DHP) Lignin With Native Lignin From Gymnosperm Wood by Thermochemolysis Using Tetramethyl-ammonium Hydroxide (TMAH). *Organic Geochemistry*, 24:593-600.

Hatcher, P.G., M.A. Nanny, R.D. Minard, S.C. Dible, and D.M. Carson. 1995. Comparison of Two Thermochemolytic Methods for the Analysis of Lignin in Decomposing Gymnosperm Wood: The CuO Oxidation Method and the Method of Thermochemolysis With Tetra-methylammonium Hydroxide (TMAH). *Organic Geochemistry*, 23:881-888.

Hatcher, P.G., D.L. VanderHart, and W.L. Earl. 1980. Use of Solid-State ^{13}C NMR in Structural Studies of Humic Acids and Humin From Holocene Sediments. *Organic Geochemistry*, 2:87-92.

Hedges, J.I., G. Eglinton, P.G Hatcher, D.L. Kirchman, C. Arnosti, S. Derenne, R.P. Evershed, I. Kogel-Knabner, J.W. de Leeuw, R. Littke, W. Michaelis, and J. Rullkotter. 2000. The Molecularly-Uncharacterized Component of Nonliving Organic Matter in Natural Environments. *Organic Geochemistry*, 31(10):945-958.

Hedges, J.I., R.G. Keil, and R. Benner. 1997. What Happens to Terrestrial Organic Matter in the Ocean? *Organic Geochemistry*, 27(5/6):195-212.

Hobbie, J.E., R.J. Daley, and S. Jasper. 1977. Use of Nuclepore Filters for Counting Bacteria by Epifluorescence Microscopy. *Applied and Environmental Microbiology*, 33:1225-1228.

Kaplan, L.A. 1992. Comparison of High-Temperature and Persulfate Oxidation Methods for Determination of Dissolved Organic Carbon in Freshwaters. *Limnology and Oceanography*, 37:1119-1125.

Kaplan, L.A., T.L. Bott, and J. Frias. 1994. Bacterial Responses to Longitudinal Nutrient Gradients Within a Bioreactor: Community and Population Phenomena. American Society for Microbiology 94th Annual Meeting. Las Vegas, Nev.

Kaplan, L.A., and T.J. Gremm. 1995a. Composition and Biodegradation Kinetics of Streamwater Dissolved Organic Matter. In *Abstracts of the American Society of Limnology and Oceanography Annual Meeting*, Reno, Nev.

Kaplan, L.A., and T.J. Gremm. 1995b. Bioavailability of Humic Substances in Stream Ecosystems. In *Abstracts of the International Humic Substances Society Meeting*, Atlanta, Ga.

Kaplan, L.A., and J.D. Newbold. 1995. Measurement of Streamwater Biodegradable Dissolved Organic Carbon With a Plug-Flow Bioreactor. *Water Research*, 29:2696-2706.

Kaplan, L.A., F. Ribas, J.C. Joret, C. Volk, J. Frias, and F. Lucena. 1996. *Measurement of Biodegradable Organic Matter With Biofilm Reactors*. Denver, Colo.: AwwaRF and AWWA.

Kaplan, L.A., and C. Volk. 1997. Specificity of Hetertrophic Communities to NOM Sources. AWWA Water Quality Technology Conference, Denver, Colo.: AWWA.

Kawakami, H. 1980. Degradation of Lignin-Related Aromatics and Lignins by Several Pseudomonads. *Lignin Biodegradation: Microbiology, Chemistry, and Potential Applications [Proc. Int. Semin.]*, 2:103-25.

Kawamura, K., R. Ishiwatari, and K. Ogura. 1987. Early Diagenesis of Organic Matter in the Water Column and Sediments: Microbial Degradation and Resynthesis of Lipids in Lake Haruna. *Organic Geochemistry*, 11:251-264.

Kurisu, F., H. Satoh, T. Mino, and T. Matsuo. 2002. Microbial Community Analysis of Thermophilic Contact Oxidation Process by Using Ribosomal RNA Approaches and the Quinone Profile Method. *Water Research*, 36:429-438.

Lee, S., and J.A. Fuhrman. 1990. DNA Hybridization to Compare Species Compositions of Natural Bacterioplankton Assemblages. *Applied and Environmental Microbiology*, 56:739-746.

Leenheer, J.A. 1994. Chemistry of Dissolved Organic Matter in Rivers, Lakes, and Reservoirs. *Advances in Chemistry Series*, 237:195-221.

Liu, W.-T., T.L. Marsh, H. Cheng, and L.J. Forney. 1997. Characterization of Microbial Diversity by Determining Terminal Restriction Fragment Length Polymorphisms of Genes Encoding 16S rRNA. *Applied and Environmental Microbiology*, 63:4516-4522.

Lucena, F., J. Frias, and F. Ribas. 1990. A New Dynamic Approach to the Determination of Biodegradable Dissolved Organic Carbon in Water. *Environmental Technology*, 12:343-347.

Mannino, A., and H.R. Harvey. 2000. Terrigenous Dissolved Organic Matter Along an Estuarine Gradient and Its Flux to the Coastal Ocean. *Organic Geochemistry*, 31:1611-1625.

McCarthy, M.D., J.I. Hedges, and R. Benner. 1998. Major Bacterial Contribution to Marine Dissolved Organic Nitrogen. *Science*, 281(5374):231-234.

McCune, B. 1999. PC-ORD. Multivariate Analysis of Ecological Data. Oregon: MJM Software Design.

McCune, B., and J.B. Grace. 2002. Analysis of Ecological Communities. Oregon: MJM Software Design.

McKinney, D.E., J.M. Bortiatynski, D.M. Carson, D.J. Clifford, J.W. De Leeuw, and P.G. Hatcher. 1996. Tetramethylammonium Hydroxide (TMAH) Thermochemolysis of the Aliphatic Biopolymer Cutan: Insights to Its Chemical Structure. *Organic Geochemistry*, 24:641-650.

Metz, G., X. Wu, and S.O. Smith. 1994. Ramped-Amplitude Cross Polarization in Magic-Angle-Spinning NMR. *Journal of Magnetic Resonance*, 110:219-227.

Meyers, P.A., and R. Ishiwatari. 1993. Lacustrine Organic Geochemistry—An Overview of Indicators of Organic Matter Sources and Diagenesis in Lake Sediments. *Organic Geochemistry*, 20(7):867-900.

Meyers, P.A., M.J. Leenheer, B.J. Eadie, and S.J. Maule. 1984. Organic Geochemistry of Suspended and Settling Particulate Matter in Lake Michigan. *Geochimica et Cosmochimica Acta*, 48:443-52.

Moeseneder, M.M., C. Winter, and G.J. Herndl. 2001. Horizontal and Vertical Complexity of Attached and Free-Living Bacteria of the Eastern Mediterranean Sea, Determined by 16S rDNA and rRNA Fingerprints. *Limnology and Oceanography*, 46:95-107.

Moll, D.M., and R.S. Summers. 1999. Assessment of Drinking Water Filter Microbial Communities Using Taxonomic and Metabolic Profiles. *Water Science and Technology*, 39(7):83-89.

Napolitano, G.E. 1999. *Fatty Acids as Trophic and Chemical Markers in Freshwater Ecosystems, Lipids in Freshwater Ecosystems*. Edited by B.C. Arts and M.T. Wainman. New York: Springer-Verlag.

Newbold, J.D., T.L. Bott, L.A. Kaplan, B.W. Sweeney, and R.L. Vannote. 1997. Organic Matter Dynamics in White Clay Creek, Pennsylvania, USA. *Journal of the North American Benthological Society*, 16:46.

Newbold, J.D., B.W. Sweeney, J.K. Jackson, and L.A. Kaplan. 1995. Concentrations and Export of Solutes From Six Mountain Streams in Northwestern Costa Rica. *Journal of the North American Benthological Society*, 14:21-37.

Pitter, P., and J. Chudoba. 1990. *Biodegradability of Organic Substances in the Aquatic Environment*. CRC Press.

Pomeroy, L.R. 1974. The Ocean's Food Web, a Changing Paradigm. *BioScience*, 24:499-504.

Prevost, M., G. Dubreuil, R. Desjardins, and R. MacLean. 1997. Bioreactors for the Rapid Determination of Biodegradable Dissolved Organic Carbon (BDOC) in Drinking Water: Feed Mode Impact. AWWA Water Quality Technology Conference, November 9-12, 1997. Denver, Colo.: AWWA.

Raskin, L., B.E. Rittmann, and D.A. Stahl. 1996. Competition and Coexistence of Sulfate-Reducing and Methanogenic Populations in Anaerobic Biofilms. *Applied and Environmental Microbiology*, 62:3847-3857.

Reichenbach H. 1992. The Order *Cytophagales*. In *The Prokaryotes*. Edited by A. Balows, H.G. Truper, M. Dworkin, W. Harder, and K.H. Schleifer. pp. 3631-3675. Springer-Verlag.

Robson, J.N., and S.J. Rowland. 1988. Biodegradation of Highly Branched Isoprenoid Hydrocarbons—A Possible Explanation of Sedimentary Abundance. *Organic Geochemistry*, 13(4-6):691-695.

Sambrook, J., E.F. Fritsch, and T. Maniatus. 1989. *Molecular Cloning: A Laboratory Manual*. Cold Spring Harbor Press.

Schmidt, T.M., E.F. Delong, and N.R. Pace. 1991. Analysis of a Marine Picoplankton Community by 16S Ribosomal RNA Gene Cloning and Sequencing. *Journal of Bacteriology*, 173:4371-4378.

Scholz, A.T., R.M. Horrall, J.C. Cooper, and A.D. Hasler. 1976. Imprinting to Chemical Cues: The Basis for Home Stream Selection in Salmon. *Science*, 192:1247-1249.

Schulten, H.R. 1999. Analytical Pyrolysis and Computational Chemistry of Aquatic Humic Substances and Dissolved Organic Matter. *Journal of Analytical and Applied Pyrolysis*, 49:385-415.

Stahl, D.A. 1995. Application of Phylogenetically Based Hybridization Probes to Microbial Ecology. *Molecular Ecology*, 4:535-542.

Sun, L., E.M. Perdue, J.L. Meyer, and J. Weis. 1997. Use of Elemental Composition to Predict Bioavailability of Dissolved Organic Matter in a Georgia River. *Limnology and Oceanography*, 42(4):714-721.

Thomas, J.D. 1997. The Role of Dissolved Organic Matter, Particularly Free Amino Acids and Humic Substances, in Freshwater Ecosystems. *Freshwater Biology*, 38(1):1-36.

Thurman, E.M., and R.L. Malcolm. 1981. Preparative Isolation of Aquatic Humic Substances. *Environmental Science and Technology*, 15:463-466.

Vallino, J.J., C.S. Hopkinson, and J.E. Hobbie. 1996. Modeling Bacterial Utilization of Dissolved Organic Matter: Optimization Replaces Monod Growth Kinetics. *Limnology and Oceanography*, 41:1591-1609.

Velji, M.I., and L.J. Albright. 1986. The Dispersion of Adhered Marine Bacteria by Pyrophosphate and Ultrasound Prior to Direct Counting. *Deuxieme Colloque International de Bacteriologie Marine*, 249-259.

Volk, C.J., C.B. Volk, and L.A. Kaplan. 1997. The Chemical Composition of Biodegradable Dissolved Organic Matter in Streamwater. *Limnology and Oceanography*, 42:39-44.

Young, C.C., R.L. Burghoff, L.G. Keim, V. Minak-Bernero, J.R. Lute, and S.M. Hinton. 1993. Polyvinylpyrrolidone-Agarose Gel Electrophoresis Purification of Polymerase Chain Reaction-Amplifiable DNA From Soils. *Applied and Environmental Microbiology*, 59:1972-1974.

Zang, X., J.D.H. van Heemst, K.J. Dria, and P.G. Hatcher. 2000. Encapsulation of Protein in Humic Acid From a Histosol as an Explanation for the Occurrence of Organic Nitrogen in Soil and Sediment. *Organic Geochemisty*, 31:679-695.

Zhou, J.B, D.S. Xia, D.S. Treves, L.–Y. Wu, T.L. Marsh, R.V. O'Neill, A.V. Paulumbo, and J.M. Tiedje. 2002. Spatial and Resource Factors Influencing High Microbial Diversity in Soil. *Applied and Environmental Microbiology*, 68:326-334.

ABBREVIATIONS

Ad	aldehyde
A_i	peak area for the analyte
A_{IS}	peak area for the internal standard
Al	alcohol
BDOC	biodegradable DOC
BOM	biodegradable organic matter
BSA	bovine serum albumin
°C	degrees Celsius
CPMAS	cross-polarization magic angle spinning
CuO	cupric oxide
c.v.	coefficient of variation
DEPC	diethyl pyrocarbamate
DNA	deoxyribonucleic acid
DOC	dissolved organic carbon
DR	Delaware River
DTCHO	dissolved total carbohydrates
EBCT	empty-bed contact time
EI	electron impact
EMC	epifluorescence microscopic counts
FAMEs	fatty acid methyl esters
fmol	femtomoles
G	guaiacyl
GC/MS	gas chromatography/mass spectroscopy
gdw	gram dry weight
Hz	hertz
i.d.	internal diameter
kHz	kilohertz
M	molar
mg/L	milligrams per liter
m/h	meters per hour
mL	milliliter
mL/min	milliliters per minute

mM	millimeter
ms	millisecond
m/z	mass/charge
μL	microliter
μm	micrometer
μs	microsecond
n	number of replicates
N	nitrogen
ng	nanogram
NIST	National Institute of Standards and Technology
NMR	nuclear magnetic resonance
NOM	natural organic matter
O	oxygen
OC	organic carbon
P	p-hydroxyphenyl
PCR	polymerase chain reaction
PLFA	phospholipid fatty acid
ppm	parts per million
PVP	polyvinlypyrrolidone
RDOM	refractory dissolved organic matter
RNA	ribosomal ribonucleic acid
rpm	revolutions per minute
RRF	relative response factor
RSF	rapid sand filter
RT	Rio Tempisquito
S	syringyl
SD	standard deviation
SR	Schuylkill River
TDS	total dissolved solids
TIC	total ion current
TMAH	tetramethylammonium hydroxide
TOC	total organic carbon
TPPM	two pulse phase modulation
t-RFLP	terminal restriction fragment length polymorphisms
U	units
UPGMA	unweighted pair-group method
V	volt
V/cm	volts per centimeter

W	watt
WCC	White Clay Creek
Wt_i	weight of analyte
Wt_{IS}	weight of the internal standard
w/v	weight per volume
x	mean
×	times

Lightning Source UK Ltd.
Milton Keynes UK
UKHW05f0504270918
329553UK00005B/804/P